Powder Surface Area and Porosity

Powder Technology Series

EDITED BY BRIAN SCARLETT

Delft University of Technology
The Netherlands

Many materials exist in the form of a disperse system, for example powders, pastes, slurries, emulsions and aerosols. The study of such systems necessarily arises in many technologies but may alternatively be regarded as a separate subject which is concerned with the manufacture, characterization and manipulation of such systems. Chapman and Hall were one of the first publishers to recognize the basic importance of the subject, going on to instigate this series of books. The series does not aspire to define and confine the subject without duplication, but rather to provide a good home for any book which has a contribution to make to the record of both the theory and the application of the subject. We hope that all engineers and scientists who concern themselves with disperse systems will use these books and that those who become expert will contribute further to the series.

Particle Size Measurement
T. Allen
4th edn, hardback (0 412 35070 X), 832 pages

Powder Surface Area and Porosity
S. Lowell and Joan E. Shields
3rd edn, hardback (0 412 39690 4), 256 pages

Pneumatic Conveying of Solids
R.D. Marcus, L.S. Leung, G.E. Klinzing and F. Rizk
Hardback (0 412 21490 3), 592 pages

Particle Technology
Hans Rumpf
Translated by F.A. Bull
Hardback (0 412 35230 3), 216 pages

Powder Surface Area and Porosity

S. LOWELL
Quantachrome Corporation, USA

JOAN E. SHIELDS
*C.W. Post Campus of
Long Island University, USA*

THIRD EDITION

CHAPMAN & HALL
London · New York · Tokyo · Melbourne · Madras

UK	Chapman & Hall, 2–6 Boundary Row, London SE1 8HN
USA	Chapman & Hall, 29 West 35th Street, New York NY10001
JAPAN	Chapman & Hall Japan, Thomson Publishing Japan, Hirakawacho Nemoto Building, 7F, 1-7-11 Hirakawa-cho, Chiyoda-ku, Tokyo 102
AUSTRALIA	Chapman & Hall Australia, Thomas Nelson Australia, 102 Dodds Street, South Melbourne, Victoria 3205
INDIA	Chapman & Hall India, R. Seshadri, 32 Second Main Road, CIT East, Madras 600 035

First edition 1979 published as *Introduction to Powder Surface Area* by John Wiley & Sons Inc., New York
Second edition 1984 published as *Powder Surface Area and Porosity* by Chapman and Hall Ltd
Third edition 1991

© 1984, 1991 S. Lowell and J.E. Shields

Typeset in 10/12pt Times by Best-set Typesetter Ltd, Hong Kong
Printed in Great Britain by T.J. Press Ltd, Padstow, Cornwall

ISBN 0 412 39690 4

Apart from any fair dealing for the purposes of research or private study, or criticism or review, as permitted under the UK Copyright Designs and Patents Act, 1988, this publication may not be reproduced, stored or transmitted, in any form or by any means, without the prior permission in writing of the publishers, or in the case of reprographic reproduction only in accordance with the terms of the licences issued by the Copyright Licensing Agency in the UK, or in accordance with the terms of licences issued by the appropriate Reproduction Rights Organization outside the UK. Enquiries concerning reproduction outside the terms stated here should be sent to the publishers at the UK address printed on this page.
The publisher makes no representation, express or implied, with regard to the accuracy of the information contained in this book and cannot accept any legal responsibility or liability for any errors or omissions that may be made.

British Library Cataloguing in Publication Data
Lowell, S. (Seymour) –1931–
 An introduction to powder characterization.
 1. Powders
 I. Title
 620.43

ISBN 0-412-39690-4

Library of Congress Cataloging-in-Publication Data
available

Contents

Preface		x
Symbols		xi
PART 1	**THEORETICAL**	1
1	**Introduction**	3
1.1	Real surfaces	3
1.2	Factors affecting surface area	3
1.3	Surface area from particle size distributions	5
2	**Gas adsorption**	7
2.1	Introduction	7
2.2	Physical and chemical adsorption	8
2.3	Physical adsorption forces	9
3	**Adsorption isotherms**	11
4	**Langmuir and BET theories (kinetic isotherms)**	14
4.1	The Langmuir isotherm, Type I	14
4.2	The Brunauer, Emmett, and Teller (BET) theory	17
4.3	Surface areas from the BET equation	22
4.4	The meaning of monolayer coverage	23
4.5	The BET constant and site occupancy	24
4.6	Applicability of the BET theory	25
4.7	Some criticism of the BET theory	27
5	**The single point BET method**	30
5.1	Derivation of the single point method	30
5.2	Comparison of the single point and multipoint methods	31
5.3	Further comparisons of the multi- and single point methods	32
6	**Adsorbate cross-sectional areas**	35
6.1	Cross-sectional areas from the liquid molar volume	35
6.2	Nitrogen as the standard adsorbate	38
6.3	Some adsorbate cross-sectional areas	41

7	**Other surface area methods**	42
7.1	Harkins and Jura relative method	42
7.2	Harkins and Jura absolute method	44
7.3	Permeametry	46
8	**Pore analysis by adsorption**	52
8.1	The Kelvin equation	52
8.2	Adsorption hysteresis	55
8.3	Types of hysteresis	56
8.4	Total pore volume	58
8.5	Pore size distributions	59
8.6	Modelless pore size analysis	65
8.7	$V-t$ curves	68
9	**Microporosity**	72
9.1	Introduction	72
9.2	Langmuir plots for microporous surface area	72
9.3	Extensions of Polanyi's theory for micropore volume and area	73
9.4	The t-method	77
9.5	The α_s-method	84
9.6	The micropore analysis method	85
9.7	Total micropore volume and surface area	88
10	**Theory of wetting and capillarity for mercury porosimetry**	90
10.1	Introduction	90
10.2	Young and Laplace equation	91
10.3	Wetting or contact angles	93
10.4	Capillarity	94
10.5	Washburn equation	96
11	**Interpretation of mercury porosimetry data**	99
11.1	Applications of the Washburn equation	99
11.2	Intrusion–extrusion curves	100
11.3	Common features of porosimetry curves	103
11.4	Solid compressibility	104
11.5	Surface area from intrusion curves	105
11.6	Pore size distribution	107
11.7	Volume ln radius distribution	110
11.8	Pore surface area distribution	110
11.9	Pore length distribution	111
11.10	Pore population	111
11.11	Plots of porosimetry functions	112
11.12	Comparisons of porosimetry and gas adsorption	119

12	**Hysteresis, entrapment and contact angle**	121
12.1	Introduction	121
12.2	Contact angle changes	123
12.3	Porosimetric work	124
12.4	Theory of porosimetry hysteresis	126
12.5	Pore potential	127
12.6	Other hysteresis theories	130
12.7	Equivalency of mercury porosimetry and gas adsorption	131
13	**Particle size**	135
13.1	Introduction	135
13.2	Stokes' law	135
13.3	X-ray absorption	137
13.4	Particle size distributions	138
13.5	Reynolds number	141
13.6	Particle porosity and density	142
13.7	Wall effects	142
13.8	Particle concentration	143
13.9	Diffusion effects	144
13.10	Terminal velocity	150
PART 2	**EXPERIMENTAL**	153
14	**Adsorption measurements — preliminaries**	155
14.1	Reference standards	155
14.2	Other preliminary precautions	156
14.3	Representative samples	156
14.4	Sample conditioning	160
15	**Vacuum volumetric measurements**	162
15.1	Nitrogen adsorption	162
15.2	Deviations from ideality	165
15.3	Sample cells	165
15.4	Evacuation and degassing	165
15.5	Temperature control	166
15.6	Isotherms	166
15.7	Low surface areas	168
15.8	Saturated vapor pressure	170
15.9	Automated instrumentation	171
15.10	Quasi–equilibrium systems	171
16	**Dynamic methods**	174
16.1	Influence of helium	174

16.2	Nelson and Eggertsen continuous flow method	175
16.3	Carrier gas and detector sensitivity	177
16.4	Design parameters for continuous flow apparatus	180
16.5	Signals and signal calibration	185
16.6	Adsorption and desorption isotherms by continuous flow	188
16.7	Low surface area measurements	189
16.8	Data reduction — continuous flow	193
16.9	Single point method	194
17	**Other flow methods**	**197**
17.1	Pressure jump method	197
17.2	Continuous isotherms	198
17.3	Frontal analysis	198
18	**Gravimetric method**	**202**
18.1	Electronic microbalances	202
18.2	Buoyancy corrections	202
18.3	Thermal transpiration	204
18.4	Other gravimetric methods	204
19	**Comparison of experimental adsorption methods**	**206**
20	**Chemisorption**	**210**
20.1	Introduction	210
20.2	Chemisorption equilibrium and kinetics	210
20.3	Chemisorption isotherms	212
20.4	Surface titrations	215
21	**Mercury porosimetry**	**217**
21.1	Introduction	217
21.2	Pressure generators	217
21.3	Dilatometer	218
21.4	Continuous scan porosimetry	218
21.5	Logarithmic signals from continuous scan porosimetry	221
21.6	Low-pressure intrusion–extrusion scans	222
21.7	Contact angle for mercury porosimetry	223
22	**Density measurement**	**227**
22.1	True density	227
22.2	Apparent density	230
22.3	Bulk density	230
22.4	Tap density	231
22.5	Effective density	231
22.6	Density by Mercury porosimetry	232

23	**Particle size analysis**	235
23.1	Introduction	235
23.2	Sample preparation	235
23.3	Sedimentation fluid	238
23.4	Dispersion	238
23.5	Data reduction	239

References 243

Index 248

Preface

The rapid growth of interest in powders and their surface properties in many diverse industries prompted the writing of this book for those who have the need to make meaningful measurements without the benefit of years of experience. It is intended as an introduction to some of the elementary theory and experimental methods used to study the surface area, porosity, density, and particle size of powders. It may be found useful by those with little or no training in solid surfaces who have the need to learn quickly the rudiments of surface area, density, pore size, and particle size measurements.

S. Lowell
J.E. Shields

Symbols

Use of symbols for purposes other than those indicated in the following table are so defined in the text. Some symbols not shown in the table are also defined in the text.

$\mathscr{A}$	adsorbate cross-sectional area
A	area; condensation coefficient; collision frequency
C	BET constant
c	concentration
D	diameter; coefficient of thermal diffusion
E	adsorption potential
f	permeability aspect factor
F	flow rate; force; feed rate
g	gravitational constant
G	Gibbs free energy
G^s	free surface energy
h	heat of immersion per unit area; height
H	enthalpy
H_i	heat of immersion
H_{sv}	heat of adsorption
i	BET intercept; filament current
k	thermal conductivity; specific reaction rate
K	Harkins–Jura constant
ℓ	length
L	heat of liquefaction
M	mass
$\bar{M}$	molecular weight
MPa	megapascals
n	number of moles
N	number of molecules; number of particles
$\bar{N}$	Avogadro's number
$\mathscr{N}$	molecular collisions per square cm per second
P	pressure
P_0	saturated vapor pressure
p	porosity
PSIA	pounds per square inch absolute

PSIG	pounds per square inch gauge
r	radius
r_k	core radius
r_p	pore radius
r_h	hydraulic radius
R	gas constant; resistance
s	BET slope
S	specific surface area; entropy
S_t	total surface area
t	time; statistical depth
T	absolute temperature
U	pore potential
v	linear flow velocity; settling velocity
V	volume
$\bar{V}$	molar volume
V_p	pore volume
W	work; weight
W_m	monolayer weight
X	mole fraction
α	temperature coefficient
β	affinity coefficient, nonideality correction; compressibility
γ	surface tension
η	viscosity
θ	contact angle
θ_0	fraction of surface unoccupied by adsorbate
θ_n	fraction of surface covered by n layers of adsorbate
μm	micrometers (10^{-6} m)
ν	vibrational frequency
π	surface pressure
ρ	density
τ	monolayer depth; time per revolution; time for one cycle
ψ	change in particle diameter per collector per revolution

Part 1
THEORETICAL

1
Introduction

1.1 REAL SURFACES

There is a convenient mathematical idealization which asserts that a cube of edge length, ℓ cm, possesses a surface area of $6\ell^2$ cm^2 and that a sphere of radius r cm exhibits $4\pi r^2$ cm^2 of surface. In reality, however, mathematical, perfect or ideal geometric forms are unattainable since under microscopic examinations all real surfaces exhibit flaws. For example, if a 'super microscope' were available one would observe surface roughness due not only to the atomic or molecular orbitals at the surface but also due to voids, steps, pores and other surface imperfections. These surface imperfections will always create real surface area greater than the corresponding theoretical area.

1.2 FACTORS AFFECTING SURFACE AREA

When a cube, real or imaginary, of one meter edge length is subdivided into smaller cubes each one micrometer (10^{-6} meter) in length there will be formed 10^{18} particles, each exposing an area of 6×10^{-12} square meters (m^2). Thus, the total area of all the particles is 6×10^6 m^2. This millionfold increase in exposed area is typical of the large surface areas exhibited by fine powders when compared to undivided material. Whenever matter is divided into smaller particles new surfaces must be produced with a corresponding increase in surface area.

In addition to particle size, the particle shape contributes to the surface area of the powder. Of all geometric forms, a sphere exhibits the minimum area-to-volume ratio while a chain of atoms, bonded only along the chain axis, will give the maximum area-to-volume ratio. All particulate matter possesses geometry and, therefore, surface areas between these two extremes. The dependence of surface area on particle shape is readily shown by considering two particles of the same composition and of equal weight, M, one particle a cube of edge length ℓ and the other spherical with radius r. Since the particle density ρ is independent of particle shape[†] one can write

---

[†] For sufficiently small particles the density can vary slightly with changes in the area to volume ratio. This is especially true if particles are ground to size and atoms near the surface are disturbed from their equilibrium position.

$$M_{cube} = M_{sphere} \tag{1.1}$$

$$(V\rho)_{cube} = (V\rho)_{sphere} \tag{1.2}$$

$$\ell^3_{cube} = \tfrac{4}{3}\pi r^3_{sphere} \tag{1.3}$$

$$\frac{S_{cube}\,\ell_{cube}}{6} = S_{sphere}\,\frac{r_{sphere}}{3} \tag{1.4}$$

$$\frac{S_{cube}}{S_{sphere}} = \frac{2r_{sphere}}{\ell_{cube}} \tag{1.5}$$

Thus, for particles of equal weight, the cubic area, will exceed the spherical area, S_{sphere}, by a factor of $2\,r/\ell$.

The range of specific surface area[†] can vary widely depending upon the particle's size and shape and also the porosity.[‡] The influence of pores can often overwhelm the size and external shape factors. For example, a powder consisting of spherical particles exhibits a total surface area, S_t, as described by equation (1.6):

$$S_t = 4\pi(r_1^2 N_1 + r_2^2 N_2 + \ldots + r_i^2 N_i) = 4\pi \sum_{i=1} r_i^2 N_i \tag{1.6}$$

where r_i and N_i are the average radii and number of particles respectively in the size range i. The volume of the same powder sample is

$$V = \tfrac{4}{3}\pi(r_1^3 N_1 + r_2^3 N_2 + \ldots + r_i^3 N_i) = \tfrac{4}{3}\pi \sum_{i=1} r_i^3 N_i \tag{1.7}$$

Replacing V in equation (1.7) by the ratio of mass to density, M/ρ, and dividing equation (1.6) by (1.7) gives the specific surface area

$$S = \frac{S_t}{M} = \frac{3\sum_i N_i r_i^2}{\rho \sum_i N_i r_i^3} \tag{1.8}$$

For spheres of uniform radius equation (1.8) becomes

$$S = \frac{3}{\rho r} \tag{1.9}$$

Thus, powders consisting of spherical particles of 0.1 micrometer (μm) radius with densities near $3\,\text{g cm}^{-3}$ will exhibit surface areas about $10^5\,\text{cm}^2\,\text{g}^{-1}$

[†] The area exposed by one gram of powder is called the 'specific surface area'.
[‡] Porosity is defined here as surface flaws which are deeper than they are wide.

($10 \, m^2 \, g^{-1}$). Similar particles with radii of $1.0 \, \mu m$ would exhibit a tenfold decrease in surface area. However, if the same $1.0 \, \mu m$ radii particles contained extensive porosity they could exhibit specific surface areas well in excess of $1000 \, m^2 \, g^{-1}$, clearly indicating the significant contribution that pores can make to the surface area.

1.3 SURFACE AREA FROM PARTICLE SIZE DISTRIBUTIONS

Although particulates can assume all regular geometric shapes and, in most instances, highly irregular shapes, most particle size measurements are based on the so called 'equivalent spherical diameter'. This is the diameter of a sphere which would behave in the same manner as the test particle being measured in the same instrument. For example, the Coulter Counter [1] is a commonly used instrument for determining particle sizes. Its operation is based on the momentary increase in the resistance of an electrolyte solution which results when a particle passes through a narrow aperture between two electrodes. The resistance change is registered in the electronics as a rapid pulse. The pulse height is proportional to the particle volume and therefore, the particles are sized as equivalent spheres.

Stokes' law [2] is another concept around which several instruments are designed to give particle size or size distributions. Stokes' law is used to determine the settling velocity of particles in a fluid medium as a function of their size. Equation (1.10) is a useful form of Stokes' law

$$D = \sqrt{\frac{18\eta v}{(\rho_s - \rho_f)g}} \tag{1.10}$$

where D is the particle diameter, η is the coefficient of viscosity, v is the settling velocity, g is the gravitational constant and ρ_s and ρ_f are the densities of the solid and the fluid, respectively. Allen [3] gives an excellent discussion of the various experimental methods associated with sedimentation size analysis and a detailed X-ray sedimentation technique is described in Chapters 13 and 23. Regardless of the experimental method employed, non-spherical particles will be measured as larger or smaller equivalent spheres depending on whether the particles settle faster or more slowly than spheres of the same mass. Modifications of Stokes' law have been used in centrifugal devices to enhance the settling rates but are subject to the same limitations of yielding only the equivalent spherical diameter.

Optical devices, based upon particle attenuation of a light beam or measurement of scattering angles and intensity, also give equivalent spherical diameters.

Permeametric methods, discussed in a later chapter, are often used to determine average particle size. The method is based upon the impedance offered to the fluid flow by a packed bed of powder. Again, equivalent spherical diameter is the calculated size.

Sieving is another technique which sizes particles according to their smallest dimension but gives no information on particle shape.

Electron microscopy can be used to estimate particle shape, at least in two dimensions. A further limitation is that only relatively few particles can be viewed.

Attempts to measure surface area based on any of the above methods will give results significantly less than the true value, in some cases by factors of 10^3 or greater depending upon particle shape, surface irregularities and porosity. At best, surface areas calculated from particle size will establish the lower limit by the implicit assumptions of sphericity or some other regular geometric shape, and by ignoring the highly irregular nature of real surfaces.

2
Gas adsorption

2.1 INTRODUCTION

Examination of powdered materials with an electron microscope can generally disclose the presence of surface imperfections and pores. However, those imperfections or irregularities smaller than the microscope's resolving power will remain hidden. Also hidden is the internal structure of the pores, their inner shape and dimensions, their volume and volume distribution as well as their contribution to the surface area. However, by enveloping each particle of a powder sample in an adsorbed film, the method of gas adsorption can probe the surface irregularities and pore interiors even at the atomic level. In this manner, a very powerful method is available which can generate detailed information about the morphology of surfaces.

To some extent, adsorption always occurs when a clean solid surface is exposed to vapor.[†] Invariably the amount adsorbed on a solid surface will depend upon the absolute temperature T, the pressure P, and the interaction potential E between the vapor (adsorbate) and the surface (adsorbent). Therefore, at some equilibrium pressure and temperature the weight W of gas adsorbed on a unit weight of adsorbent is given by

$$W = F(P, T, E) \tag{2.1}$$

Usually the quantity adsorbed is measured at constant temperature and equation (2.1) reduces to

$$W = F(P, E) \tag{2.2}$$

A plot of W versus P, at constant T, is referred to as the sorption isotherm of a particular vapor solid interface. Were it not for the fact that E, the interaction potential, varies with the properties of the vapor and the solid and

---

[†] A vapor is defined here as a gas below its critical temperature and, therefore, condensable.

also changes with the extent of adsorption, all adsorption isotherms would be identical.

2.2 PHYSICAL AND CHEMICAL ADSORPTION

Depending upon the strength of the interaction, all adsorption processes can be divided into the two categories of chemical and physical adsorption. The former, also called irreversible adsorption or chemisorption, is characterized mainly by large interaction potentials, which lead to high heats of adsorption often approaching the value of chemical bonds. This fact, coupled with other spectroscopic, electron spin resonance, and magnetic susceptibility measurements confirms that chemisorption involves true chemical bonding of the gas or vapor with the surface. Because chemisorption occurs through chemical bonding it is often found to occur at temperatures above the critical temperature of the adsorbate. Strong bonding to the surface is necessary in the presence of higher thermal energies, if adsorption is to occur at all. Also, as is true for most chemical reactions, chemisorption is usually associated with an activation energy. In addition, chemisorption is necessarily restricted to, at most, a single layer of chemically bound adsorbate on the surface. Another important factor relating to chemisorption is that the adsorbed molecules are more localized on the surface when compared to physical adsorption. Because of the formation of a chemical bond between an adsorbate molecule and a specific site on the surface the adsorbate is less free to migrate about the surface. This fact often enables the number of active sites on catalysts to be determined by simply measuring the quantity of chemisorbed gas.

The second category, reversible or physical adsorption, exhibits characteristics that make it most suitable for surface area determinations as indicated by the following:

1. Physical adsorption is accompanied by low heats of adsorption with no violent or disruptive-structural changes occurring to the surface during the adsorption measurement.
2. Unlike chemisorption, physical adsorption may lead to surface coverage by more than one layer of adsorbate. Thus, pores can be filled by the adsorbate for pore volume measurements.
3. At elevated temperatures, physical adsorption does not occur or is sufficiently slight that relatively clean surfaces can be prepared on which to make accurate surface area measurements.
4. Physical adsorption equilibrium is achieved rapidly since no activation energy is required as is generally true in chemisorption. An exception here is adsorption in small pores where diffusion can limit the adsorption rate.
5. Physical adsorption is fully reversible, enabling both the adsorption and desorption processes to be studied.

6. Physically adsorbed molecules are not restricted to specific sites and are free to cover the entire surface. For this reason, surface areas rather than number of sites can be calculated.

2.3 PHYSICAL ADSORPTION FORCES

Upon adsorption, the entropy change of the adsorbate, ΔS_a, is necessarily negative since the condensed state is more ordered than the gaseous state because of the loss of at least one degree of translational freedom. A reasonable assumption for physical adsorption is that the entropy of the adsorbent remains essentially constant and certainly does not increase by more than the adsorbate's entropy decreases. Therefore, ΔS for the entire system is necessarily negative. The spontaneity of the adsorption process requires that the Gibbs free energy, ΔG, also be a negative quantity. Based upon the entropy and free energy changes, the enthalpy change, ΔH, accompanying physical adsorption is always negative, indicating an exothermic process, as shown by equation (2.3)

$$\Delta H = \Delta G + T \Delta S \tag{2.3}$$

An important interaction at the gas-solid interface during physical adsorption is due to dispersion forces. These forces are present regardless of the nature of other interactions and often account for the major part of the adsorbate–adsorbent potential. The nature of dispersion forces was first recognized in 1930 by F. London [4] who postulated that the electron motion in an atom or molecule would lead to a rapidly oscillating dipole moment. At any instant, the lack of symmetry of the electron distribution about the nuclei imparts a transient dipole moment to an atom or molecule, which vanishes when averaged over a longer time interval. When in close proximity, the rapidly oscillating dipoles of neighboring molecules couple into phase with each other leading to a net attracting potential. This phenomenon is associated with the molecular dispersion of light due to the light's electromagnetic field interaction with the oscillating dipole.

Among other adsorbate–adsorbent interactions contributing to adsorption are:

1. Ion-dipole — an ionic solid and electrically neutral but polar adsorbate.
2. Ion-induced dipole — a polar solid and polarizable adsorbate.
3. Dipole-dipole — a polar solid and polar adsorbate.
4. Quadrapole interactions — symmetrical molecules with atoms of different electronegativities, such as CO_2 possess no dipole moment but do have a quadrapole ($^-O - CC^{++} - O^-$), which can lead to interactions with polar surfaces.

It is evident from the above that adsorption forces are similar in nature and origin to the forces that lead to liquefaction of vapors and that the same intermolecular interactions are responsible for both phenomena. Thus, those vapors with high boiling points and, therefore, strong intermolecular interactions will also tend to be strongly adsorbed.

Above the critical temperature, the thermal energy possessed by gas molecules is sufficient to overcome the forces leading to liquefaction. Because of the similarity between adsorption and liquefaction, the critical temperature can be used as an estimate of the maximum temperature at which significantly measurable amounts of physical adsorption can occur.

3
Adsorption isotherms

Brunauer, Deming, Deming and Teller [5], based upon an extensive literature survey, found that all adsorption isotherms fit into one of the five types shown below in Fig. 3.1.

The five isotherm shapes depicted in Fig. 3.1 each reflects some unique condition. Each of these five isotherms and the conditions leading to its occurrence are discussed below.

Type I isotherms are encountered when adsorption is limited to, at most, only a few molecular layers. This condition is encountered in chemisorption where the asymptotic approach to a limiting quantity indicates that all of the surface sites are occupied. In the case of physical adsorption, type I isotherms are encountered with microporous powders whose pore size does not exceed a few adsorbate molecular diameters. A gas molecule, when inside pores of these small dimensions, encounters the overlapping potential from the pore walls which enhances the quantity of gas adsorbed at low relative pressures. At higher pressures, the pores are filled by adsorbed or condensed adsorbate leading to the plateau, indicating little or no additional adsorption after the micropores have been filled. Physical adsorption that produces the Type I isotherm indicates that the pores are microporous and that the exposed surface resides almost exclusively within the micropores, which once filled with adsorbate, leave little or no external surface for additional adsorption.

Type II isotherms are most frequently encountered when adsorption occurs on nonporous powders or on powders with pore diameters larger than micropores. The inflection point or knee of the isotherm usually occurs near the completion of the first adsorbed monolayer and with increasing relative pressure, second and higher layers are completed until at saturation the number of adsorbed layers becomes infinite.

Type III isotherms are characterized principally by heats of adsorption which are less than the adsorbate heat of liquefaction. Thus, as adsorption proceeds, additional adsorption is facilitated because the adsorbate interaction with an adsorbed layer is greater than the interaction with the adsorbent surface.

12 Adsorption isotherms

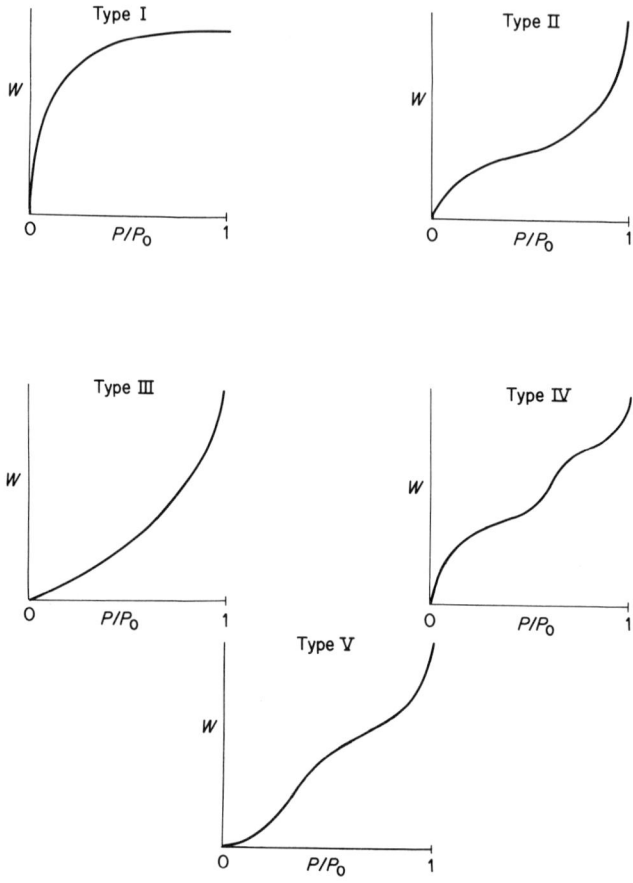

Figure 3.1 The five isotherm classifications according to BDDT [5]. W, weight adsorbed; P, adsorbate equilibrium pressure; P_0, adsorbate saturated equilibrium vapor pressure; P/P_0, relative pressure. Condensation occurs at $P/P_0 \geq 1$.

Type IV isotherms occur on porous adsorbents possessing pores in the radius range of approximately 15–1000 angstroms (Å). The slope increase at higher elevated pressures indicates an increased uptake of adsorbate as the pores are being filled. As is true for the Type II isotherms, the knee of the Type IV isotherm generally occurs near the completion of the first monolayer.

Type V isotherms result from small adsorbate–adsorbent interaction potentials similar to the Type III isotherms. However, Type V isotherms are also associated with pores in the same range as those of the Type IV isotherms.

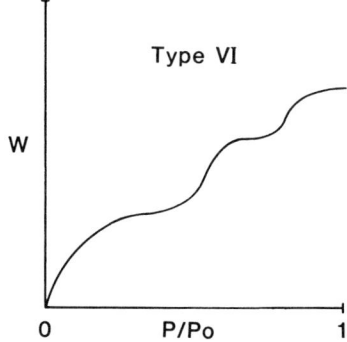

Figure 3.2 Type VI isotherm.

A new rare type of isotherm recently has been found. This type VI isotherm, illustrated in Fig. 3.2, exhibits a series of steps.

4
Langmuir and BET theories (kinetic isotherms)

The success of kinetic theories directed toward the measurements of surface areas depends upon their ability to predict the number of adsorbate molecules required exactly to cover the solid with a single molecular layer. Equally important is the cross-sectional area of each molecule or the effective area covered by each adsorbed molecule on the surface. The surface area, then, is the product of the number of molecules in a completed monolayer and the effective cross-sectional area of an adsorbate molecule. The number of molecules required for the completion of a monolayer will be considered in this chapter and the adsorbate cross-sectional area will be discussed in Chapter 6.

4.1 THE LANGMUIR ISOTHERM, TYPE I

The asymptotic approach of the quantity adsorbed toward a limiting value indicates that Type I isotherms are limited to, at most, a few molecular layers.[†] In the case of chemisorption, only one layer can be bonded to the surface and, therefore, chemisorption always exhibits a Type I isotherm.[‡] Although it is possible to calculate the number of molecules in the monolayer from the Type I chemisorption isotherm, some serious difficulty is encountered when attempts are made to apply the cross-sectional adsorbate area. This difficulty arises because chemisorption tightly binds and localizes the adsorbate to a specific surface site so that the spacing between adsorbed molecules will depend upon the adsorbent surface structures as well as the size of the adsorbed molecules or atoms. In those cases where the surface sites are widely separated, the calculated surface area will be smaller than the actual

---

[†] Physical adsorption on microporous materials shows Type I isotherms because the pores limit adsorption to only a few molecular layers. Once the micropores are filled there is only a small fraction of the original surface exposed for continued adsorption.
[‡] Under certain conditions physical adsorption can occur on top of a chemisorbed layer but this does not change the essential point being made here.

The Langmuir isotherm, Type I

value because the number of molecules in the monolayer will be less than the maximum number which the surface can accommodate. Nevertheless, it will be instructive to consider the Type I isotherm in preparation for the more rigorous requirements of the other four types.

Using a kinetic approach, Langmuir [6] was able to describe the Type I isotherm with the assumption that adsorption was limited to a monolayer. According to the kinetic theory of gases, the number of molecules $\mathcal{N}$ striking each cm^2 of surface per second is given by

$$\mathcal{N} = \frac{\bar{N}P}{(2\pi\bar{M}RT)^{\frac{1}{2}}} \qquad (4.1)$$

where $\bar{N}$ is Avogadro's number, P is the adsorbate pressure, $\bar{M}$ is the adsorbate molecular weight, R is the gas constant and T is the absolute temperature. If θ_0 is the fraction of the surface unoccupied (i.e., with no adsorbed molecules) then the number of collisions with bare or uncovered surface per square centimeter of surface each second is

$$\mathcal{N}' = kP\theta_0 \qquad (4.2)$$

The constant k is $\bar{N}/(2\pi\bar{M}RT)^{\frac{1}{2}}$. The number of molecules striking and adhering to each square centimeter of surface is

$$\mathcal{N}_{ads} = kP\theta_0 A_1 \qquad (4.3)$$

where A_1 is the condensation coefficient and represents the probability of a molecule being adsorbed upon collision with the surface.

The rate at which adsorbed molecules leave each square centimeter of surface is given by

$$\mathcal{N}_{des} = N_m \theta_1 v_1 e^{-E/RT} \qquad (4.4)$$

where N_m is the number of adsorbate molecules in a completed monolayer of one square centimeter, θ_1 is the fraction of the surface occupied by the adsorbed molecules, E is the energy of adsorption and v_1 is the vibrational frequency of the adsorbate normal to the surface when adsorbed. Actually, the product $N_m \theta_1$ is the number of molecules adsorbed per square centimeter. Multiplication by v_1 converts this number of molecules to the maximum rate at which they can leave the surface. The term $e^{-E/RT}$ represents the probability that an adsorbed molecule possesses adequate energy to overcome the net attractive potential of the surface. Thus, equation (4.4) contains all the parameters required to describe the rate at which molecules leave each square centimeter of surface.

16 Langmuir and BET theories (kinetic isotherms)

At equilibrium the rates of adsorption and desorption are equal. Thus, equating (4.3) and (4.4):

$$N_m \theta_1 v_1 e^{-E/RT} = kP\theta_0 A_1 \tag{4.5}$$

Recognizing that $\theta_0 = 1 - \theta_1$, yields

$$N_m \theta_1 v_1 e^{-E/RT} = kPA_1 - \theta_1 kPA_1 \tag{4.6}$$

then

$$\theta_1 = \frac{kPA_1}{N_m v_1 e^{-E/RT} + kPA_1} \tag{4.7}$$

Allowing

$$K = \frac{kA_1}{N_m v_1 e^{-E/RT}} \tag{4.8}$$

Substitution of equation (4.8) into (4.7) gives

$$\theta_1 = \frac{KP}{1 + KP} \tag{4.9}$$

The assumption implicit in equation (4.8) is that the adsorption energy E is constant, which implies an energetically uniform surface.

Up to and including one layer of coverage one can write

$$\theta_1 = \frac{N}{N_m} = \frac{W}{W_m} \tag{4.10}$$

where N and N_m are the number of molecules in the incompleted and completed monolayer, respectively, and W/W_m is the weight adsorbed relative to the weight adsorbed in a completed monolayer. Substituting W/W_m for θ_1 in equation (4.9) yields

$$\frac{W}{W_m} = \frac{KP}{1 + KP} \tag{4.11}$$

Equation (4.11) is the Langmuir equation for Type I isotherms. Rearrangement of equation (4.11) gives

$$\frac{P}{W} = \frac{1}{KW_m} + \frac{P}{W_m} \tag{4.12}$$

A plot of P/W versus P will give a straight line of slope $1/W_m$ and intercept $1/KW_m$ from which both K and W_m can be calculated.

Having established W_m, the sample surface area S_t can then be calculated from equation (4.13)

$$S_t = N_m \mathscr{A} = \frac{W_m \bar{N} \mathscr{A}}{\bar{M}} \quad (4.13)$$

where $\mathscr{A}$ and $\bar{M}$ are the cross-sectional area and the molecular weight of the adsorbate, and $\bar{N}$ is Avogadro's number.

Although the Langmuir equation describes Type I and sometimes chemisorption isotherms, it fails to be adequately general to treat physical adsorption and the Type II–Type V isotherms. In addition, surface area measurements obtained from Type I isotherms are subject to uncertainties, regardless of whether chemisorption or physical adsorption is occurring. In chemisorption, localization of the adsorbate molecules leaves the value of $\mathscr{A}$ seriously in question, since the adsorbate will adsorb only at active surface sites, leaving an unspecified area around each chemisorbed molecule. When applied to physical adsorption, the Type I isotherm is associated with condensation in micropores with no clearly defined region of monolayer coverage.

4.2 THE BRUNAUER, EMMETT, AND TELLER (BET) THEORY [7]

During the process of physical adsorption, at very low relative pressure, the first sites to be covered are the more energetic ones. Those sites with higher energy on a chemically pure surface reside within narrow pores where the pore walls provide overlapping potentials. Other high-energy sites lie between the horizontal and vertical edges of surface steps where the adsorbate can interact with surface atoms in two planes. In general, wherever the adsorbate is afforded the opportunity to interact with overlapping potentials or an increased number of surface atoms there will be a higher energy site. On surfaces consisting of heteroatoms, such as organic solids or impure materials, there will be variations in adsorption potential depending upon the nature of the atoms of functional groups exposed at the surface.

That the more energetic sites are covered first as the pressure is increased does not imply that no adsorption occurs on sites of less potential. Rather, it implies that the average residence time of a physically-adsorbed molecule is longer on the higher-energy sites. Accordingly, as the adsorbate pressure is allowed to increase, the surface becomes progressively coated and the probability increases that a gas molecule will strike and be adsorbed on a previously bound molecule. Clearly then, prior to complete surface coverage the formation of second and higher adsorbed layers will commence. In reality, there exists no pressure at which the surface is covered with exactly a completed physically adsorbed monolayer. The effectiveness of the Brunauer, Emmett and Teller (BET) theory is that it enables an experimental determination of the number of molecules required to form a monolayer despite the fact that exactly one monomolecular layer is never actually formed.

Brunauer, Emmett, and Teller, in 1938, extended Langmuir's kinetic theory to multilayer adsorption. The BET theory assumes that the uppermost molecules in adsorbed stacks are in dynamic equilibrium with the vapor. This means that, where the surface is covered with only one layer of adsorbate, an equilibrium exists between that layer and the vapor, and where two layers are adsorbed, the upper layer is in equilibrium with the vapor, and so forth. Since the equilibrium is dynamic, the actual location of the surface sites covered by one, two or more layers may vary but the number of molecules in each layer will remain constant.

Using the Langmuir theory and equation (4.5) as a starting point to describe the equilibrium between the vapor and the adsorbate in the first layer

$$N_m \theta_1 v_1 e^{-E_1/RT} = kP\theta_0 A_1 \qquad \text{(cf.4.5)}$$

By analogy, for the fraction of surface covered by two layers one may write

$$N_m \theta_2 v_2 e^{-E_2/RT} = kP\theta_1 A_2 \qquad (4.14)$$

In general, for the nth layer one obtains

$$N_m \theta_n v_n e^{-E_n/RT} = kP\theta_{n-1} A_n \qquad (4.15)$$

The BET theory assumes that the terms v, E, and A remain constant for the second and higher layers. This assumption is justifiable only on the grounds that the second and higher layers are all equivalent to the liquid state. This undoubtedly approaches reality as the layers proceed away from the surface but is somewhat questionable for the layers nearer the surface because of polarizing forces. Nevertheless, using this assumption one can write a series of equations, using L as the heat of liquefaction

$$N_m \theta_1 v_1 e^{-E_1/RT} = kP\theta_0 A_1 \qquad \text{(cf.4.5)}$$

$$N_m \theta_2 v e^{-L/RT} = kP\theta_1 A \qquad (4.16a)$$

$$N_m \theta_3 v e^{-L/RT} = kP\theta_2 A \qquad (4.16b)$$

and, in general, for the second and higher layers

The Brunauer, Emmett, and Teller (BET) theory

$$N_m \theta_n v e^{-L/RT} = k P \theta_{n-1} A \tag{4.16c}$$

From these it follows that

$$\frac{\theta_1}{\theta_0} = \frac{kPA_1}{N_m v_1 e^{-E_1/RT}} = \alpha \tag{4.17a}$$

$$\frac{\theta_2}{\theta_1} = \frac{kPA}{N_m v e^{-L/RT}} = \beta \tag{4.17b}$$

$$\frac{\theta_3}{\theta_2} = \frac{kPA}{N_m v e^{-L/RT}} = \beta \tag{4.17c}$$

$$\frac{\theta_n}{\theta_{n-1}} = \frac{kPA}{N_m v e^{-L/RT}} = \beta \tag{4.17d}$$

then

$$\theta_1 = \alpha \theta_0 \tag{4.18a}$$

$$\theta_2 = \beta \theta_1 = \alpha \beta \theta_0 \tag{4.18b}$$

$$\theta_3 = \beta \theta_2 = \alpha \beta^2 \theta_0 \tag{4.18c}$$

$$\theta_n = \beta \theta_{n-1} = \alpha \beta^{n-1} \theta_0 \tag{4.18d}$$

The total number of molecules adsorbed at equilibrium is

$$N = N_m \theta_1 + 2 N_m \theta_2 + \ldots + n N_m \theta_n = N_m (\theta_1 + 2\theta_2 + \ldots + n\theta_n) \tag{4.19}$$

Substituting for $\theta_1, \theta_2, \ldots$ from equations (4.18 a–d) gives

$$\begin{aligned} \frac{N}{N_m} &= \alpha \theta_0 + 2\alpha \beta \theta_0 + 3\alpha \beta^2 \theta_0 + \ldots + n\alpha \beta^{n-1} \theta_0 \\ &= \alpha \theta_0 (1 + 2\beta + 3\beta^2 + \ldots + n\beta^{n-1}) \end{aligned} \tag{4.20}$$

Since both α and β are assumed to be constants, one can write

$$\alpha = C\beta \tag{4.21}$$

This defines C by using equations (4.17a) and (4.17b–d) as

20 Langmuir and BET theories (kinetic isotherms)

$$\frac{A_1 v_2}{A_2 v_1} e^{(E_i - L)/RT} = C \tag{4.22}$$

Substituting $C\beta$ for α in equation (4.20) yields

$$\frac{N}{N_m} = C\theta_0(\beta + 2\beta^2 + 3\beta^3 + \ldots + n\beta^n) \tag{4.23}$$

The preceding summation is just $\beta/(1-\beta)^2$. Therefore,

$$\frac{N}{N_m} = \frac{C\theta_0 \beta}{(1-\beta)^2} \tag{4.24}$$

Necessarily

$$1 = \theta_0 + \theta_1 + \theta_2 + \ldots + \theta_n \tag{4.25}$$

then

$$\theta_0 = 1 - (\theta_1 + \theta_2 + \ldots + \theta_n) = 1 - \sum_{n=1}^{\infty} \theta_n \tag{4.26}$$

Substituting equation (4.26) into (4.24) gives

$$\frac{N}{N_m} = \frac{C\beta}{(1-\beta)^2} \left(1 - \sum_{n=1}^{\infty} \theta_n \right) \tag{4.27}$$

Replacing θ_n in equation (4.27) with $\alpha \beta^{n-1} \theta_0$ from equation (4.18d) yields

$$\frac{N}{N_m} = \frac{C\beta}{(1-\beta)^2} \left(1 - \alpha \theta_0 \sum_{n=1}^{\infty} \beta^{n-1} \right) \tag{4.28}$$

and introducing $C\beta$ from equation (4.21) in place of α gives

$$\frac{N}{N_m} = \frac{C\beta}{(1-\beta)^2} \left(1 - C\theta_0 \sum_{n=1}^{\infty} \beta^n \right) \tag{4.29}$$

The summation in equation (4.29) is

$$\sum_{n=1}^{\infty} \beta^n = \beta + \beta^2 + \ldots + \beta^n = \frac{\beta}{1-\beta} \tag{4.30}$$

Then

$$\frac{N}{N_m} = \frac{C\beta}{(1-\beta)^2}\left(1 - C\theta_0\frac{\beta}{1-\beta}\right) \tag{4.31}$$

From equation (4.24) we have

$$\frac{C\beta}{(1-\beta)^2} = \frac{N}{N_m}\frac{1}{\theta_0} \tag{cf.4.24}$$

Then equation (4.31) becomes

$$1 = \frac{1}{\theta_0}\left(1 - C\theta_0\frac{\beta}{1-\beta}\right) \tag{4.32}$$

and

$$\theta_0 = \frac{1}{1 + C\beta/(1-\beta)} \tag{4.33}$$

Introducing θ_0 from equation (4.33) into (4.24) yields

$$\frac{N}{N_m} = \frac{C\beta}{(1-\beta)(1-\beta+C\beta)} \tag{4.34}$$

When β equals unity, N/N_m becomes infinite. This can physically occur when adsorbate condenses on the surface or when $P/P_0 = 1$.

Rewriting equation (4.17d) for $P = P_0$, gives

$$1 = \frac{kAP_0}{N_m v e^{-L/RT}} \tag{4.35}$$

but

$$\beta = \frac{kAP}{N_m v e^{-L/RT}} \tag{cf.4.17d}$$

then

$$\beta = \frac{P}{P_0} \tag{4.36}$$

Introducing this value for β into (4.34) gives

$$\frac{N}{N_m} = \frac{C(P/P_0)}{(1 - P/P_0)[1 - P/P_0 + C(P/P_0)]} \quad (4.37)$$

Recalling that $N/N_m = W/W_m$ (equation 4.10) and rearranging equation (4.37) gives the BET equation in final form,

$$\frac{1}{W[(P_0/P) - 1]} = \frac{1}{W_m C} + \frac{C - 1}{W_m C}\left(\frac{P}{P_0}\right) \quad (4.38)$$

If adsorption occurs in pores limiting the number of layers then the summation in equation (4.27) is limited to n and the BET equation takes the form

$$\frac{W}{W_m} = \frac{C}{[(P_0/P) - 1]} \frac{[1 - (n+1)(P/P_0)^n + n(P/P_0)^{n+1}]}{[1 + (C-1)P/P_0 - C(P/P_0)^{n+1}]} \quad (4.39)$$

Equation (4.39) reduces to (4.38) with $n = \infty$ and to the Langmuir equation (4.11) with $n = 1$.

4.3 SURFACE AREAS FROM THE BET EQUATION

The determination of surface areas from the BET theory is a straight forward application of equation (4.38). A plot of $1/W[(P_0/P) - 1]$ versus P/P_0, as shown in Fig. 4.1, will yield a straight line usually in the range $0.05 \leq P/P_0 \leq 0.35$.

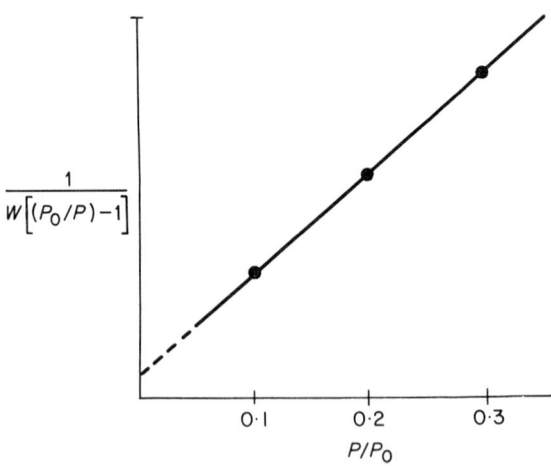

Figure 4.1 Typical BET plot.

The slope s and the intercept i of a BET plot are

$$s = \frac{C-1}{W_m C} \tag{4.40}$$

$$i = \frac{1}{W_m C} \tag{4.41}$$

Solving the preceding equations for W_m, the weight adsorbed in a monolayer, gives

$$W_m = \frac{1}{s+i} \tag{4.42}$$

and the solution for C, the BET constant, gives

$$C = \frac{s}{i} + 1 \tag{4.43}$$

The total surface area can be calculated from equation (4.13), viz.,

$$S_t = \frac{W_m \bar{N} \mathscr{A}}{\bar{M}} \tag{cf.4.13}$$

and the specific surface area can be determined by dividing S_t by the sample weight.

4.4 THE MEANING OF MONOLAYER COVERAGE

Hill [8] has shown that when sufficient adsorption has occurred to cover the surface with exactly one layer of molecules, the fraction of surface, $(\theta_0)_m$, not covered by any molecule is dependent on the BET C value and is given by

$$(\theta_0)_m = \frac{C^{\frac{1}{2}} - 1}{C - 1} \tag{4.44}$$

It is evident from equation (4.44) that when sufficient adsorption has occurred to form a monolayer there is still always some fraction of surface unoccupied. Indeed, only for C values approaching infinity will θ_0 approach zero and in such cases the high adsorbate-surface interaction can only result from chemisorption. For nominal C values, say near 100, the fraction of surface unoccupied, when exactly sufficient adsorption has occurred to form a monolayer, is 0.091. Therefore, on the average each occupied site contains

24 Langmuir and BET theories (kinetic isotherms)

about 1.1 molecules. The implication here is that the BET equation indicates the weight of adsorbate required to form a single molecular layer on the surface, although no such phenomenon as a uniform monolayer exists in the case of physical adsorption.

4.5 THE BET CONSTANT AND SITE OCCUPANCY

Equation (4.44) is used to calculate the fraction of surface unoccupied when $W = W_m$, that is, when just a sufficient number of molecules have been adsorbed to give monolayer coverage. Lowell [9] has derived an equation that can be used to calculate the fraction of surface covered by adsorbed molecules of one or more layers in depth. Lowell's equation is

$$(\theta_i)_m = C \left(\frac{C^{\frac{1}{2}} - 1}{C - 1} \right)^{i+1} \tag{4.45}$$

where θ_i represents the fraction of surface covered by layers i molecules deep. The subscript m denotes that equation (4.45) is valid only when sufficient adsorption has occurred to make $W = W_m$.

Table 4.1 shows the fraction of surface covered by layers of various depth, as calculated from equation (4.44) for $i = 0$ and (4.45) for $i \neq 0$, as a function of the BET C value.

Table 4.1 Values for $(\theta_i)_m$ from equations (4.44) and (4.45)

i	$C = 1000$	$C = 100$	$C = 10$	$C = 1$
0	0.0307	0.0909	0.2403	0.5000
1	0.9396	0.8264	0.5772	0.2500
2	0.0288	0.0751	0.1387	0.1250
3	0.0009	0.0068	0.0333	0.0625
4		0.0006	0.0080	0.0313
5		0.0001	0.0019	0.0156
6			0.0005	0.0078
7			0.0001	0.0039
8				0.0019
9				0.0009
10				0.0005
11				0.0002
12				0.0001

For the special case of $C = 1$, $(\theta_i)_m$ is evaluated by

$$\lim_{C \to 1} \frac{C^{\frac{1}{2}} - 1}{C - 1} = 0.5$$

Equations (4.44) and (4.45) should not be taken to mean that the adsorbate is necessarily arranged in neat stacks of various heights but rather as an indication of the fraction of surface covered with the equivalent of i molecules regardless of their specific arrangement, lateral mobility, and equilibrium with the vapor phase.

Of further interest is the fact that when the BET equation (4.38) is solved for the relative pressure corresponding to monomolecular coverage, ($W = W_m$), one obtains

$$\left(\frac{P}{P_0}\right)_m = \frac{C^{\frac{1}{2}} - 1}{C - 1} \tag{4.46}$$

The subscript m above refers to monolayer coverage. Equating (4.46) and (4.44) produces the interesting fact that

$$\theta_0 = \left(\frac{P}{P_0}\right)_m \tag{4.47}$$

That is, the numerical value of the relative pressure required to make W equal to W_m is also the fraction of surface unoccupied by adsorbate.

4.6 APPLICABILITY OF THE BET THEORY

Although derived over fifty years ago, the BET theory continues to be almost universally used because of its simplicity, its definitiveness, and its ability to accommodate each of the five isotherm types. The mathematical nature of the BET equation in its most general form, equation (4.39), gives the Langmuir or Type I isotherm when $n = 1$. Plots of W/W_m versus P/P_0 using equation (4.38) conform to Type II or Type III isotherms for C values greater than and less than 2, respectively. Fig. 4.2 shows the shape of several isotherms for various values of C. The data for Fig. 4.2 are shown in Table 4.2 with values of W/W_m calculated from equation (4.38) after rearrangement to

$$\frac{W_m}{W} = \left(1 - \frac{P}{P_0}\right) + \frac{1}{C}\left(\frac{P}{P_0} + \frac{P_0}{P} - 2\right) \tag{4.48}$$

The remaining two isotherms, Types IV and V, are modifications of the Type II and Type III isotherms due to the presence of pores.

Rarely, if ever, does the BET theory exactly match an experimental isotherm over its entire range of relative pressures. In a qualitative sense, however, it does provide a theoretical foundation for the various isotherm shapes. Of equal significance is the fact that in the region of relative pressures near completed monolayers ($0.05 \leq P/P_0 \leq 0.35$) the BET theory and

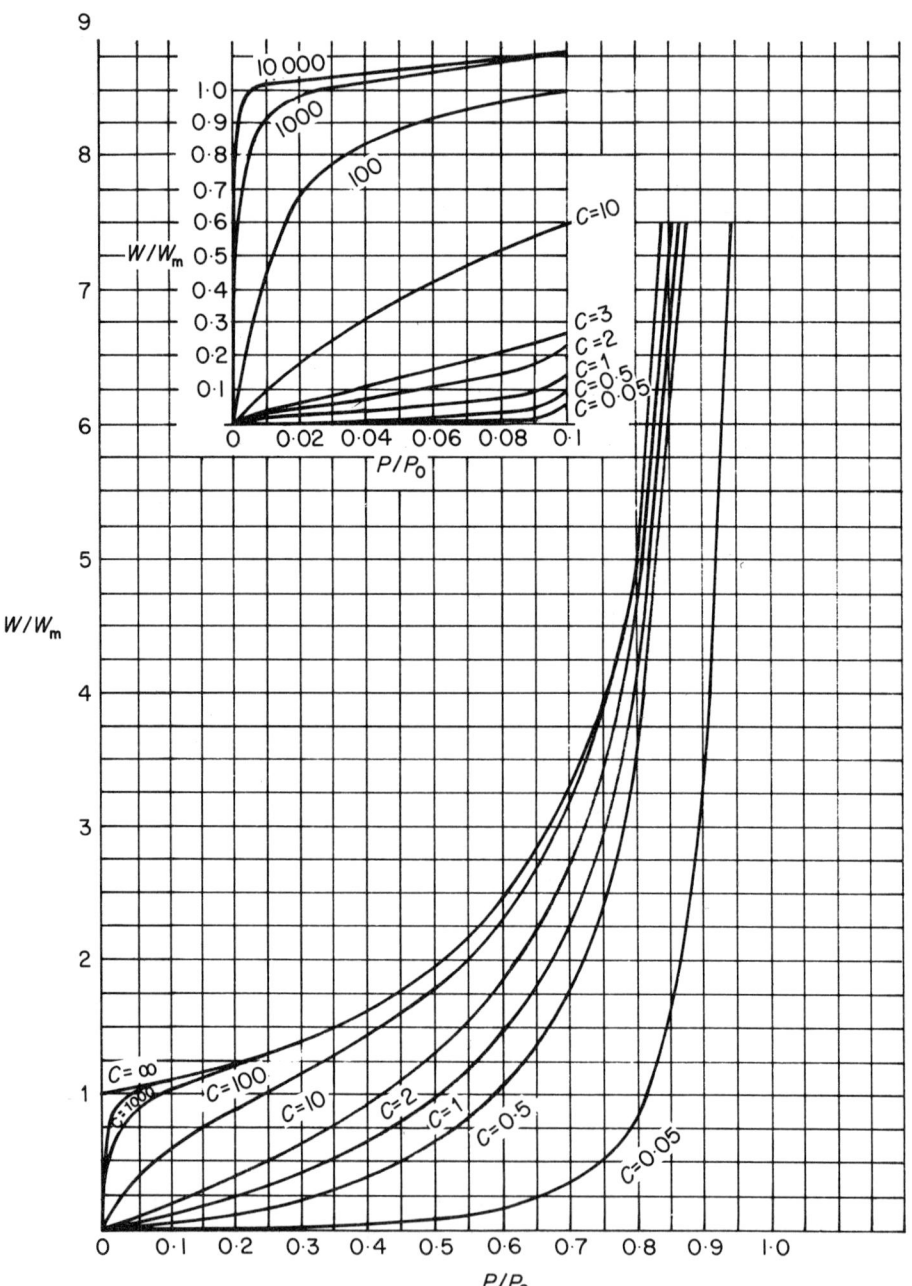

Figure 4.2 Isotherm shapes as a function of BET C values.

experimental isotherms do agree very well, leading to a powerful and extremely useful method of surface area determination.

The fact that most monolayers are completed in the range $0.05 \leq P/P_0 \leq 0.35$ reflects the value of most C constants. As shown in Table 4.2, the value of W/W_m equals unity in the previous range of relative pressures for C values between 3 and 1000, which covers the great majority of all isotherms.

The sparsity of data regarding Type III isotherms, with C values of 2 or less, leaves open the question of the usefulness of the BET method for determining surface areas when Type III isotherms are encountered. Often in this case it is possible to change the adsorbate to one with a higher C value, thereby changing the isotherm shape. Brunauer, Copeland, and Kantro [10], however, point to considerable success in calculating the surface area from Type III isotherms as well as predicting the temperature coefficient of the same isotherms.

4.7 SOME CRITICISM OF THE BET THEORY

In spite of the success of the BET theory, some of the assumptions upon which it is founded are not above criticism. One questionable assumption is that of an energetically homogeneous surface, that is, all the adsorption sites are energetically identical. Further, the BET model ignores the influence of lateral adsorbate interactions.

Brunauer [11] answers these criticisms by pointing out that lateral interaction between adsorbate molecules necessarily increases as the surface becomes more completely covered. The interaction with the surface, however, decreases with increasing adsorption up to monolayer coverage since on an energetically heterogeneous surface the high energy sites will be occupied

Table 4.2 Values of W/W_m and relative pressures for various values of C

P/P_0	$C = 0.05$	$C = 0.5$	$C = 1$	$C = 2$	$C = 3$	$C = 10$	$C = 100$	$C = 1000$
0.02	0.001	0.010	0.020	0.040	0.059	0.173	0.685	0.973
0.05	0.003	0.027	0.052	0.100	0.143	0.362	0.884	1.03
0.10	0.006	0.058	0.111	0.202	0.278	0.585	1.02	1.10
0.20	0.015	0.139	0.250	0.417	0.536	0.893	1.20	1.25
0.30	0.030	0.253	0.429	0.660	0.804	1.16	1.40	1.43
0.40	0.054	0.417	0.667	0.952	1.11	1.45	1.64	1.66
0.50	0.095	0.667	1.00	1.33	1.50	1.82	1.98	2.00
0.60	0.172	1.06	1.49	1.87	2.04	2.34	2.48	2.50
0.70	0.345	1.79	2.33	2.74	2.91	3.19	3.32	3.33
0.80	0.833	3.33	4.00	4.44	4.62	4.88	4.99	5.00
0.90	3.33	8.33	9.09	9.52	9.68	9.90	9.99	10.0
0.94	7.35	14.8	15.7	16.2	16.3	16.6	16.7	16.7

at lower relative pressures with occupancy of the lower energy sites occurring nearer to completion of the monolayer.

Fig. 4.3 illustrates how the lateral interactions and the surface interactions can sum to a nearly constant overall adsorption energy up to completion of the monolayer, an implicit assumption of the BET theory. This results in a constant C value from equation (4.22).

$$C = \frac{v_2 A_1}{v_1 A_2} e^{(E-L)/RT} \qquad \text{(cf.4.22)}$$

The dotted lines in Fig. 4.3 indicate the influence of very high adsorption potentials which can account, at least in part, for the failure of BET plots at very low relative pressures ($P/P_0 < 0.05$).

A further criticism of the BET theory is the assumption that the heat of adsorption of the second and higher layers is equal to the heat of liquefaction. It seems reasonable to expect that polarization forces would induce a higher heat of adsorption in the second layer than in the third, and so forth. Only after several layers are adsorbed should the heat of adsorption equal the heat of liquefaction. It is, therefore, difficult to resolve a model of molecules adsorbed in stacks while postulating that all layers above the first are thermodynamically a true liquid structure. The apparent validity of these criticisms contributes to the failure of the BET equation at high relative pressures (P/P_0

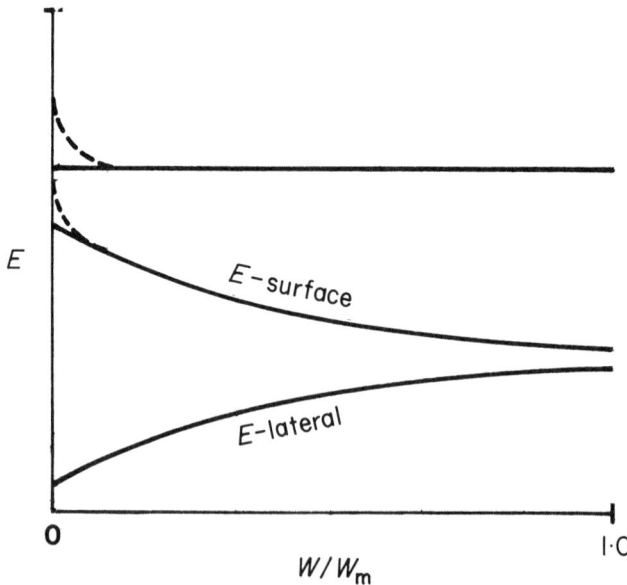

Figure 4.3 Variation in adsorption and lateral interaction potentials.

> 0.35). However, in the range of relative pressure leading to coverage near $W/W_m = 1$, the BET C values usually give heats of adsorption that are reasonable. Thus, for the great majority of isotherms the range of relative pressures between 0.05 and 0.35, the linear BET range, apparently represents a condition in which the very high energy sites have been occupied and extensive multilayer adsorption has not yet commenced. It is within these limits that the BET theory is generally valid.

Instances have been found in which BET plots were noted to be linear to relative pressures as high as 0.5, and in other cases the linear range is found only below relative pressures of 0.1 [12]. The extent to which these deviations from the usual range of linearity reflect unusual surface properties is difficult to ascertain. For example, micropores exhibit unusually high adsorption potentials due to the overlapping potential from the walls of the micropore. Under these enhanced conditions, condensation can occur within the pores at relative pressures less than 0.1, and linear BET plots are found at even lower relative pressures.

5
The single point BET method [13]

The BET theory requires that a plot of $1/W[(P_0/P) - 1]$ versus P/P_0 be linear with a finite intercept [see equation (4.38) and Fig. 4.1]. By reducing the experimental requirement to only one data point, the single point method offers the advantages of simplicity and speed often with little loss in accuracy.

5.1 DERIVATION OF THE SINGLE POINT METHOD

The slope s and intercept i of a BET plot are

$$s = \frac{C - 1}{W_m C} \tag{cf.4.40}$$

$$i = \frac{1}{W_m C} \tag{cf.4.41}$$

Then

$$\frac{s}{i} = C - 1 \tag{5.1}$$

For reasonably high values of C the intercept is small compared to the slope and in many instances may be taken as zero. With this approximation, equation (4.38), the BET equation, becomes

$$\frac{1}{W[(P_0/P) - 1]} = \frac{C - 1}{W_m C} \left(\frac{P}{P_0}\right) \tag{5.2}$$

Since $1/W_m C$, the intercept, is assumed to vanish, equation (5.2) reduces to

$$W_m = W(1 - P/P_0) \tag{5.3}$$

The total surface area as measured by the single point method, is then calculated as:

$$S_t = W\left(1 - \frac{P}{P_0}\right)\frac{\bar{N}}{\bar{M}}\mathscr{A} \tag{5.4}$$

5.2 COMPARISON OF THE SINGLE POINT AND MULTIPOINT METHODS

The error introduced by the single point method can be evaluated by examining the difference between W_m as determined by equations (5.3) and (4.38), the BET equation. Solving equation (4.38) for W_m gives

$$W_m = \left(\frac{P_0}{P} - 1\right)\left[\frac{1}{C} + \frac{C-1}{C}\left(\frac{P}{P_0}\right)\right] \tag{5.5}$$

Subtracting equation (5.3) and (5.5) and dividing by equation (5.5) gives the relative error associated with the single point method, that is,

$$\frac{(W_m)_{mp} - (W_m)_{sp}}{(W_m)_{mp}} = \frac{1 - P/P_0}{1 + (C-1)P/P_0} \tag{5.6}$$

The subscripts mp and sp refer to the multi- and single point methods, respectively.

Table 5.1 shows the relative error of the single point method compared to the multipoint method as a function of P/P_0 as calculated from equation (5.6).

The last column of Table 5.1 is established by substituting equation (4.46) into equation (5.6) for the special case when $P/P_0 = (P/P_0)_m$, thus

$$\frac{(W_m)_{mp} - (W_m)_{sp}}{(W_m)_{mp}} = \frac{C^{\frac{1}{2}} - 1}{C - 1} = \left(\frac{P}{P_0}\right)_m = \theta_0 \tag{5.7}$$

Table 5.1 Relative error using the single point method at various relative pressures

C	$P/P_0 = 0.1$	$P/P_0 = 0.2$	$P/P_0 = 0.3$	$P/P_0 = (P/P_0)_m^\dagger$
1	0.90	0.80	0.70	0.50
10	0.47	0.29	0.19	0.24
50	0.17	0.07	0.04	0.12
100	0.08	0.04	0.02	0.09
1000	0.009	0.004	0.002	0.03

$^\dagger (P/P_0)_m$ is the relative pressure which gives monolayer coverage according to a multipoint determination.

The surprising relationship above shows that when a single point analysis is made using the relative pressure which would give monolayer coverage according to the multipoint theory, the relative error will be equal to the relative pressure employed and also, according to equation (4.44), will be equal to the fraction of surface unoccupied.

A more explicit insight into the mathematical differences of the multi- and single point methods is obtained by considering a single point analysis using a relative pressure of 0.3 with a corresponding multipoint C value of 100.

From equation (5.3), the single point BET equation, one obtains

$$(W_m)_{sp} = 0.7 W_{0.3} \tag{5.8}$$

The term $(W_m)_{sp}$ refers to the monolayer weight as determined by the single point method and $W_{0.3}$ is the experimental weight adsorbed at a relative pressure of 0.3.

From equation (4.46) the relative pressure required for monolayer coverage is

$$\left(\frac{P}{P_0}\right)_m = \frac{100^{\frac{1}{2}} - 1}{100 - 1} = 0.0909 \tag{5.9}$$

Using equation (5.5), the multipoint equation, to find W_m gives

$$W_m = W_{0.3}(3.33 - 1)\left[\frac{1}{100} + \frac{100 - 1}{100}(0.3)\right] = 0.715 W_{0.3} \tag{5.10}$$

Comparison of equations (5.10) and (5.8) shows that the difference between the single and multipoint methods is identical to that shown in Table 5.1 for $P/P_0 = 0.3$ and $C = 100$; viz.,

$$\frac{0.715 - 0.700}{0.715} = 0.02 \tag{5.11}$$

5.3 FURTHER COMPARISONS OF THE MULTI- AND SINGLE POINT METHODS

The above analysis discloses that, when the BET C value is 100, the single point method using a relative pressure more than three times that required for monolayer coverage causes an error of only 2%. To further understand the BET equation and the relationship between the C value and the single point error it is useful to rewrite equation (4.38), the BET equation, as

$$\frac{W}{W_m} = \frac{C(P/P_0)}{[1 + (C - 1)P/P_0](1 - P/P_0)} \tag{5.12}$$

Using the method of partial fractions the right side of equation (5.12) can be written as

$$\frac{C(P/P_0)}{[1 + (C-1)P/P_0](1 - P/P_0)} = \frac{X}{(1 - P/P_0)} - \frac{Z}{1 + (C-1)P/P_0} \quad (5.13)$$

Recognizing that $X = Z = 1$ is a solution, gives

$$\frac{W}{W_m} = \frac{1}{1 - P/P_0} - \frac{1}{1 + (C-1)P/P_0} \quad (5.14)$$

Equation (5.14) is the BET equation expressed as the difference between two rectangular hyperbola. If the value of C is taken as infinity then equation (5.14) immediately reduces to equation (5.3), the single point BET equation. The hyperbola referred to above are shown below in Fig. 5.1.

As indicated in Fig. 5.1, curve Y, the BET curve for an arbitrary C value, approaches curve X, the single point curve, as the values of C increases. In the limiting case of C equal to ∞ the BET curve is coincident with the single

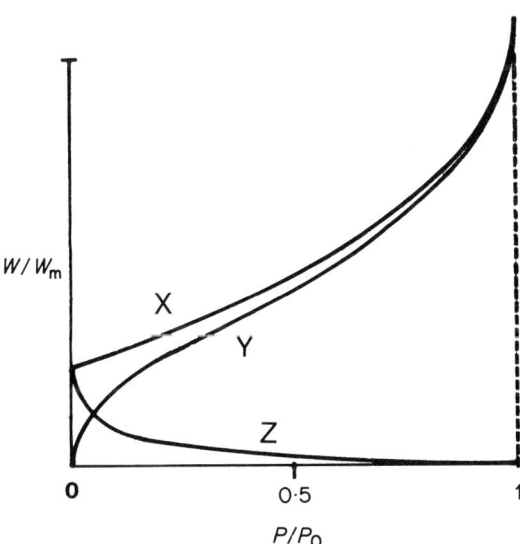

Figure 5.1 Plot of the hyperbola from equation (5.14) using an arbitrary C value. Curve $X = \dfrac{1}{1 - P/P_0}$; curve $Z = \dfrac{1}{1 + (C-1)P/P_0}$; curve $Y = X - Z$

point curve. For all other C values, the single point curve lies above the BET curve and their difference vanishes as the relative pressure approaches unity. Thus, as the value of C increases, the knee of the isotherm becomes sharper and moves toward lower relative pressures. For lower C values, curves X and Y diverge and higher relative pressures must be used to make single point surface areas conform to those obtained by the multipoint method.

The extent of divergence of curves X and Y is controlled, in a mathematical sense, by the second term in equation (5.13). It is this term which contains the C value.

Table 5.1 indicates that, regardless of the C value, the relative error is reduced by using higher relative pressures. Similarly, Fig. 5.1 shows that at sufficiently high relative pressures the BET curve and the single point curve merge regardless of the C value. It would appear that all single point analyses should be performed at the highest possible relative pressures. Although theoretically sound, the use of relative pressures above 0.3 can lead to serious errors on the large number of samples that contain pores. In a later chapter, the influence of pores is discussed, but here it is sufficient to note that once condensation in pores commences, the BET equation, which deals only with adsorption, fails adequately to describe the state of the system. Ample evidence is available to indicate that many adsorbents possess pores which cause condensation at relative pressures as small as 0.3 and in some cases at even lower values [14]. Therefore, relative pressures of 0.3 may be considered adequately high to give good agreement with multipoint measurements on most surfaces while avoiding condensation in all but microporous samples.

When used for quality control, the error associated with the single point method can be eliminated or greatly reduced if an initial multipoint analysis is performed to obtain the correct C value. Then equation (5.6) can be used to correct the results. Even an approximate value of C can be used to estimate the single point error. However, on the great majority of surfaces, the C value is sufficiently high to reduce the single point error to less than 5%.

6
Adsorbate cross-sectional areas

Using the BET equation to determine W_m, the monolayer weight, and with reasonable estimates of the adsorbate cross-sectional area, $\mathscr{A}$, the total sample surface area, S_t, in square meters, can be calculated from equation (4.13).

$$S_t = \frac{W_m \bar{N} \mathscr{A}}{\bar{M}} \times 10^{-20} \text{ m}^2 \qquad \text{(cf.4.13)}$$

with W_m in grams, $\bar{M}$ is the adsorbate molecular weight, $\bar{N}$ is Avogadro's number (6.02×10^{23} molecules per mole) and $\mathscr{A}$ in square ångströms per molecule. Division by the sample weight converts S_t to S, the specific surface area.

6.1 CROSS-SECTIONAL AREAS FROM THE LIQUID MOLAR VOLUME

A reasonable approximation of the cross-sectional area of adsorbates was proposed by Emmett and Brunauer [15]. They assumed the adsorbate molecules to be spherical and using the bulk liquid properties they calculated the cross-sectional area from

$$\mathscr{A} = \left(\frac{\bar{V}}{\bar{N}}\right)^{\frac{2}{3}} \times 10^{16} \text{ Å}^2 \qquad (6.1)$$

where $\bar{V}$ is the liquid molar volume. Equation (6.1) must be amended to reflect the molecular packing on the surface. Assuming that the liquid is structured as spheres with 12 nearest neighbors, 6 on a plane, in the usual close packed hexagonal arrangement as shown in Fig. 6.1 and that the adsorbate has the same structure on the adsorbent surface, equation (6.1) becomes

$$\mathscr{A} = 1.091 \left(\frac{\bar{V}}{\bar{N}}\right)^{\frac{2}{3}} \times 10^{16} \text{ Å}^2 \qquad (6.2)$$

36 Adsorbate cross-sectional areas

The factor 1.091 in equation (6.2) arises from the characteristics of close packed hexagonal structures. If D is the distance between centers of adjacent spheres, the spacing between the centers of adjacent rows in a plane is $3^{1/2}D/2$. The spacing between centers of adjacent planes is $(2/3)^{1/2}D$ [16]. Allowing N_x and N_y to represent the number of spheres along the X and Y axes of a plane of spheres, the planar area, A_p, is given by

$$A_p = \frac{3^{\frac{1}{2}}}{2} D^2 N_x N_y \tag{6.3}$$

If N_z is the number of planes or layers, then the volume, V, containing $N_x N_y N_z$ spheres is given by

$$V = \frac{3^{\frac{1}{2}}}{2}\left(\frac{2}{3}\right)^{\frac{1}{2}} D^3 N_x N_y N_z \tag{6.4}$$

Since $N_x N_y N_z$ represents the total number of spheres, N, in the volume, V, equation (6.4) can be expressed as

$$V = \frac{3^{\frac{1}{2}}}{2}\left(\frac{2}{3}\right)^{\frac{1}{2}} D^3 N \tag{6.5}$$

Then

$$\left(\frac{V}{N}\right)^{\frac{2}{3}} = \frac{3^{\frac{1}{3}}}{2^{\frac{2}{3}}}\left(\frac{2}{3}\right)^{\frac{1}{3}} D^2 \tag{6.6}$$

Substituting for D^2 into equation (6.3) gives

$$A_p = 1.091\, N_x N_y \left(\frac{V}{N}\right)^{\frac{2}{3}} \tag{6.7}$$

The molecular cross-sectional area $\mathscr{A}$ then can be obtained by dividing the planar area, A_p, by $N_x N_y$, the number of molecules in a plane. Thus, by dividing both numerator and denominator of the fraction, V/N, by the number of moles, yields

$$\mathscr{A} = 1.091 \left(\frac{V}{N}\right)^{\frac{2}{3}} \times 10^{16}\ \text{Å}^2 \tag{cf. 6.2}$$

That the adsorbate resides on the adsorbent surface with a structure similar to a plane of molecules within the bulk liquid, as depicted in Fig. 6.1, is a

Cross-sectional areas from the liquid molar volume 37

Figure 6.1 Sectional view of a close-packed hexagonal arrangement of spheres.

simplified view of the real situation on surfaces. Factors that make this model and therefore equation (6.2) of limited value include the following:

1. Any lateral motion of the adsorbate will be highly disruptive of any definite arrangement of adsorbate on the surface.
2. Complex molecules which rotate about several bond axes can undergo conformational changes on various surfaces and thereby exhibit different cross-sectional areas.
3. Orientation of polar molecules produces different surface arrangements depending on the polarity of the adsorbent.
4. Strong interactions with the surface lead to localized adsorption which constrains the adsorbate to a specific site. The effective adsorbate cross-sectional area will then reflect the spacing between sites rather than the actual adsorbate dimensions.
5. Fine pores may not be accessible to the adsorbate, so that a substantial portion of the surface is inaccessible to measurement. This would be particularly true for large adsorbate molecules.

The conclusion, based on the above factors, is that surface areas calculated from equation (4.13) usually give different results depending upon the adsorbate used. If the cross-sectional areas are arbitrarily revised to give surface area conformity on one sample, the revised values generally will not give surface area agreement when the adsorbent is changed.

6.2 NITROGEN AS THE STANDARD ADSORBATE

For surface area determinations the ideal adsorbate should exhibit BET C values sufficiently low to preclude localized adsorption. When the adsorbate is so strongly tied to the surface as to be constrained to specific adsorption sites, the adsorbate cross-sectional area will be determined more by the adsorbent lattice structure than by the adsorbate dimensions. This type of 'epitaxial' adsorption will lead to decreasing measured surface areas relative to the true BET value as the surface sites become more widely spaced.

On the other hand, if the C value is sufficiently small, lateral mobility of the adsorbate on the surface will tend to disrupt any tendency for the development of an organized structure and the adsorbed layer might appear more as a two dimensional gas.

In either of the preceding cases, very high or very low C values, any attempt to calculate the effective adsorbate cross-sectional areas from the bulk liquid properties will be subject to considerable error. Nitrogen, as an adsorbate, exhibits the unusual property that on almost all surfaces its C value is sufficiently small to prevent localized adsorption and yet adequately large to prevent the adsorbed layer from behaving as a two dimensional gas.

Kiselev and Eltekov [17] established that the BET C value influences the adsorbate cross-sectional area. They measured the surface area of a number of adsorbents using nitrogen. When the surface areas of the same adsorbents were measured using n-pentane as the adsorbate the cross-sectional areas of n-pentane had to be revised in order to match the surface areas measured using nitrogen. It was found that the revised areas increased hyperbolically as the n-pentane C value decreased, as shown in Fig. 6.2.

A similar relationship between n-butane cross-sectional areas and the BET C constant has been reported [18]. A plot of the revised cross-sectional areas of n-butane versus the BET constant is shown in Fig. 6.3.

Figures 6.2 and 6.3 show that plots of $(C^{1/2} - 1)/(C - 1)$ versus C give hyperbola which also match the cross-sectional area data. A plausible explanation for the observation that the fraction of surface not covered by adsorbate, $(\theta_0)_m$, increases at low C values, leading to high apparent cross-sectional areas, is that the two hydrocarbons used as adsorbates interact weakly with the adsorbent, and thus behave as two-dimensional gases on the surface. Therefore, their cross-sectional areas may reflect the area swept out by the adsorbate during their residence time on the surface rather than their actual cross-sectional areas.

In those instances of very high C values, the fraction of surface uncovered by adsorbate again increases, as a result of epitaxial deposition on specific surface sites, which when widely spaced, would lead to high apparent cross-sectional areas.

A complete plot of cross-sectional area versus the BET C value would then

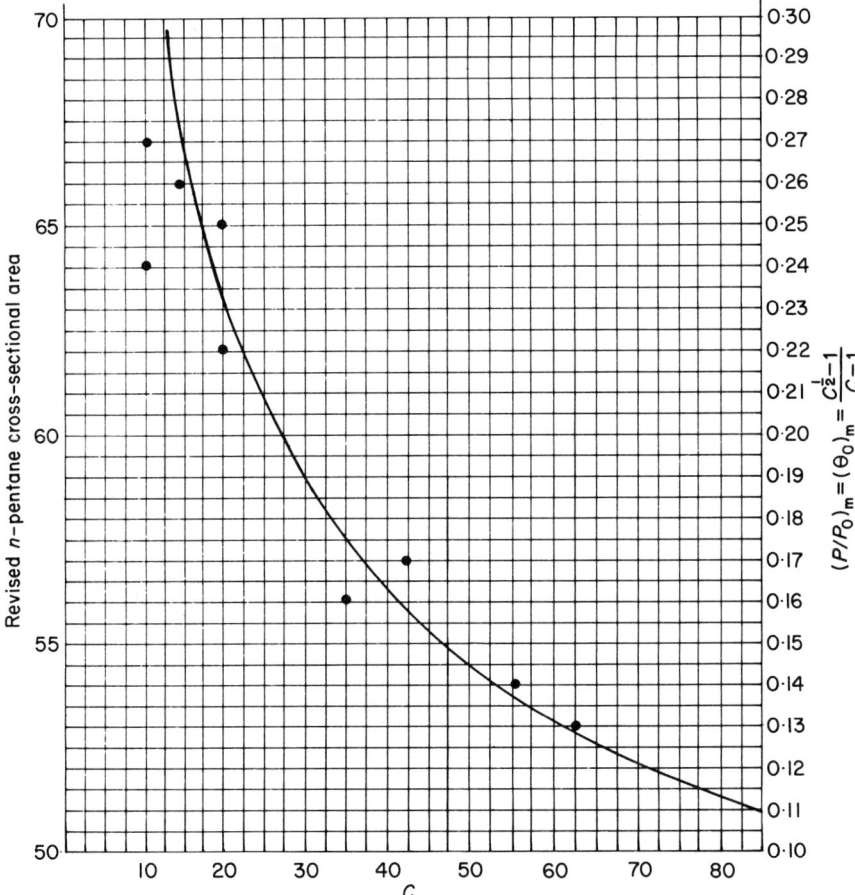

Figure 6.2 Variation of *n*-pentane cross-sectional area with the BET *C* constant (points) and $(\theta_0)_m$ (solid line).

be parabolic in shape, with the most suitable values of cross-sectional areas lying near the minimum of the parabola. For the great majority of adsorbents, the *C* constant for nitrogen lies in the range from about 50 to 300. Interactions leading to *C* values as low as 10 or 20 are not found with nitrogen nor is nitrogen chemisorbed, which would lead to adsorption on specific sites. Thus, nitrogen is uniquely suited as a desirable adsorbate, since its *C* value is not found at the extremes at each end of the parabola.

Since $(C^{1/2} - 1)/(C - 1)$, $(\theta_0)_m$, can be calculated from a BET plot, there exists a potential means of predicting the cross-sectional area variation relative to nitrogen. On surfaces that contain extensive porosity, which

40 Adsorbate cross-sectional areas

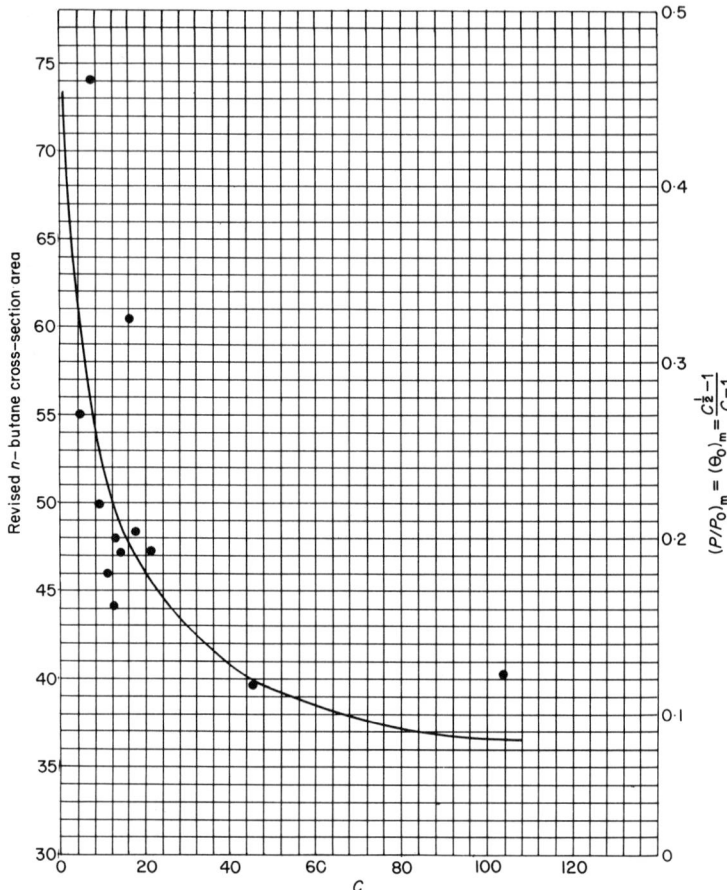

Figure 6.3 Variation of n-butane cross-sectional area with the BET C constant (points) and $(\theta_0)_m$ (solid line).

exclude larger adsorbate molecules from some pores while admitting smaller ones, it becomes even more difficult to predict any variation in the adsorbate cross-sectional area by comparison to a standard [19, 25].

In summary, it may be concluded that the uncertainty in calculating absolute cross-sectional areas, the variation in cross-sectional areas with the BET C value and the fact that on porous surfaces less area is available for larger adsorbate molecules all point to the need for a universal, although possibly arbitrary, standard adsorbate. The unique properties of nitrogen have led to its acceptance in this role with an assigned cross-sectional area of 16.2 Å² at its boiling point of $-195.6\,°C$.

Table 6.1 Approximate cross-sectional areas

Adsorbate	Area (Å^2)	Reference
argon	14.2	21
ammonia	14.6	20
benzene	40.0	22
carbon dioxide	19.5	20
carbon monoxide	16.3	20
ethane	20.5	23
krypton	19.5	24
n-butane	46.9	25
n-hexane	51.0	22
nitrogen	16.2	26
oxygen	14.1	27
water	10.8	20
xenon	2.5	28

6.3 SOME ADSORBATE CROSS-SECTIONAL AREAS

Livingston [20] has assigned values of molecular cross-sectional areas to over twenty adsorbates based on best values. These were obtained by assigning values to each adsorbate, as required, in order to make measured surface areas agree with the nitrogen value. For purposes of comparison, Livingston assigned nitrogen the value of 15.4 Å^2 and placed the greatest weight on those adsorbents which exhibited no porosity.

Table 6.1 lists some approximate adsorbate cross-sectional areas. Because the adsorbates listed are used at various temperatures and on vastly different adsorbents the values are only approximate.

With regard to cross-sectional areas, it must be kept in mind that the area occupied by a molecule or atom can often be many times its true area and the terms 'effective area' or 'occupied area' are more appropriate and possibly less misleading.

7
Other surface area methods

Because of its simplicity and straightforward applicability, the BET theory is almost universally employed for surface area measurements. However, other methods and theoretical models have been developed which are briefly outlined in this chapter. No attempt is made to derive and discuss these alternate methods completely but rather to present their essential features and to indicate how they may be used to calculate surface areas.

7.1 HARKINS AND JURA RELATIVE METHOD [29]

When a thin film of fatty acid is spread on the surface of water the surface tension of the water is reduced from γ_0 to γ. A barrier placed between pure water and water with a surface film will experience a pressure difference resulting from the tendency of the film to spread. This 'surface pressure', π, is given by

$$\pi = \gamma_0 - \gamma \tag{7.1}$$

Langmuir [30], in 1917, constructed the 'film balance' for the measurement of the 'surface' or 'spreading' pressure. Thus, it became possible to experimentally observe that adsorbed films pass through several states of molecular arrangement [31]. The various states resemble that of a two-dimensional gas, a low density liquid, and finally a higher density or condensed liquid state. In the latter case, the spreading pressure can be described by the linear relationship,

$$\pi = \alpha - \beta \mathscr{A} \tag{7.2}$$

where $\mathscr{A}$ is the effective cross-sectional area of the adsorbed molecules and α and β are constants related to the film's compressibility.

A fundamental equation derived by Gibbs [32] is used to calculate the spreading pressure of films on solids where, unlike films on liquids, it cannot be determined experimentally. Guggenheim and Adam [33] reduced Gibbs' general adsorption equation to equation (7.3) for the special case of gas adsorption.

$$d\pi = \frac{RTW}{\overline{M}S_t} d\ln P \tag{7.3}$$

The terms in equation (7.3) have previously been defined as W = weight of adsorbate, $\overline{M}$ = adsorbate molecular weight, S_t = solid surface area and P = equilibrium pressure.

Differentiating equation (7.2) and substituting for $d\pi$ in equation (7.3) yields

$$-\beta d\mathscr{A} = \frac{RTW}{\overline{M}S_t} d\ln P \tag{7.4}$$

The implicit assumption made in deriving equation (7.4) is that an adsorbed gas on a solid surface behaves similarly to a thin film of fatty acid on the surface of water.

Rewriting equation (4.13) as

$$\mathscr{A} = \frac{S_t \overline{M}}{W \overline{N}} \tag{cf.4.13}$$

and differentiating $\mathscr{A}$ with respect to W yields

$$d\mathscr{A} = -\left(\frac{S_t \overline{M}}{\overline{N}}\right) \frac{dW}{W^2} \tag{7.5}$$

Replacing $d\mathscr{A}$ in equation (7.4) by equation (7.5) and rearranging terms gives

$$d\ln P = \left(\frac{\beta \overline{M}^2 S_t^2}{RT\overline{N}}\right) \frac{dW}{W^3} \tag{7.6}$$

Integrating equation (7.6) and subtracting $\ln P_0$ from each side yields

$$\ln\left(\frac{P}{P_0}\right) = -\left(\frac{\beta \overline{M}^2 S_t^2}{2RT\overline{N}}\right) \frac{1}{W^2} + \text{const} - \ln P_0 \tag{7.7}$$

or

$$\ln\left(\frac{P}{P_0}\right) = A - \frac{B}{W^2} \qquad (7.7a)$$

According to equation (7.7a), the Harkins–Jura equation, a plot of $\ln(P/P_0)$ versus $1/W^2$ should give a straight line with a slope equal to $-B$ and an intercept equal to A. The surface area is then calculated as

$$S_t = \frac{1}{\bar{M}}\left(\frac{2\bar{N}RT}{\beta}\right)^{\frac{1}{2}} B^{\frac{1}{2}} \qquad (7.8)$$

or

$$S_t = KB^{\frac{1}{2}} \qquad (7.8a)$$

where

$$K = \frac{1}{\bar{M}}\left(\frac{2\bar{N}RT}{\beta}\right)^{\frac{1}{2}} \qquad (7.9)$$

The term K in equation (7.9) is the Harkins–Jura (HJ) constant and is assumed to be independent of the adsorbent and dependent only on the adsorbate.

In some instances, Harkins and Jura found two or more linear regions of different slopes when $\ln(P/P_0)$ is plotted versus W^{-2}. This indicates the existence of two or more liquid condensed states in which different molecular packing occurs. When this situation appears, the slope that gives best agreement with an alternate method, for example, the BET method, must be chosen. Alternatively, the temperature or the adsorbate can be changed to eliminate the ambiguity.

Emmett [34] has shown that linear HJ plots are obtained for relative pressures between 0.05 and about 0.30, the usual BET range, when the BET C value lies between 50 and 250. For C values below 10, the linear HJ region lies above relative pressures of 0.4 and for $C = 1000$ the HJ plot is linear between 0.01 and 0.18. Further, Emmett has found that the nitrogen cross-sectional area had to be adjusted from 13.6 Å^2 to 18.6 Å^2 as C varied between 50 and 250 in order to give the same surface area as the HJ method using 4.06 as the value for K.

7.2 HARKINS AND JURA ABSOLUTE METHOD [35]

In addition to the relative method, Harkins and Jura have also developed an absolute method for surface area measurement which is independent of the adsorption isotherm and is based entirely upon calorimetric data. Consider a system in which one gram of solid and α moles of vapor are transformed into a state in which the solid is immersed in a liquid made by condensing the vapor. As shown in Fig. 7.1, this transformation can be accomplished along two possible paths.

The terms ΔH_i, L, ΔH_{SV} and i used in Fig. 7.1 are all enthalphy changes defined as follows: ΔH_i is the heat of immersion of the solid into the liquid, L is the latent heat of condensation, ΔH_{SV} is the heat of adsorption when the solid is equilibrated with saturated vapor, and i is the heat liberated when solid in equilibrium with saturated vapor is immersed into liquid. Using Hess's law of heat summation

$$\Delta H_1 + \Delta H_2 = \Delta H_3 + \Delta H_4 + \Delta H_5 \tag{7.10}$$

then

$$\Delta H_i = (\Delta H_{SV} - \alpha_1 L) + i \tag{7.11}$$

The quantity $(\Delta H_{SV} - \alpha_1 L)$ is the integral heat of adsorption. This value as well as the value of i can be measured calorimetrically. The value of i is actually zero if the isotherm approaches the ordinate asymptotically. If the isotherm cuts the ordinate at a finite angle, i will be finite but small.

For powder samples that differ only in their surface area, the heat of immersion will be proportional to the surface area. Thus,

$$H_i = h_i S \tag{7.12}$$

where h_i is the heat of immersion per unit area of solid.

Gregg and Sing [36] have tabulated values of h_i for various solids and liquids. However, values of h_i found in the literature must be used with great

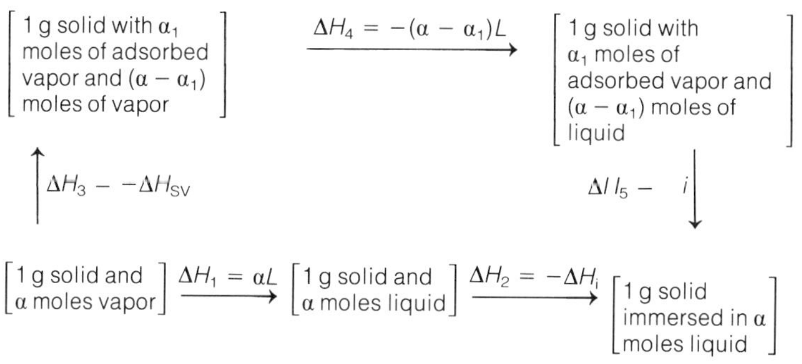

Figure 7.1

caution since an error of even one calorie would produce a surface area error of 42 m² with a typical h_i value of 100 erg/cm². Serious errors can also be encountered if close attention is not paid to the liquid purity since the measured heat may reflect a quantity of strongly adsorbed impurities. With regard to the solid itself, even identical chemical analyses will not guarantee that two samples will have the same h_i value. The same material prepared differently or with different histories can possess varying amounts of lattice strain which will produce meaningful differences in h_i values. Samples with differing pore sizes but otherwise identical will also exhibit varying h_i values due to the variation in the potential fields within the pores. Ideally therefore, it is desirable to work with carefully annealed nonporous materials when using the heat of immersion method. These types of samples are rarely encountered and their low surface areas will yield small heat values, making the experimental determination of their surface areas a tedious and difficult procedure. These difficulties indicate clearly that the immersion method is of questionable value as a rapid or routine method for surface area measurements.

An important contribution made by the Harkins and Jura absolute method however, must not be overlooked. Their measurements of some specific surface areas give confirmation to the value of 16.2 Å² for the cross-sectional area of nitrogen. This value, when employed with the BET theory, gave exactly the same specific surface area as the HJ absolute method.

7.3 PERMEAMETRY

Consider a gas, near ambient pressure and temperature, forced by a small pressure gradient to flow through the channels of a packed bed of powder. At room temperature, a gas molecule can be adsorbed on a solid surface for an extremely short time but not less than the time required for one vibrational cycle or about 10^{-13} seconds. When the adsorbed molecule leaves the surface it will, on the average, have a zero velocity component in the direction of flow. After undergoing one or several gas phase collisions, it will soon acquire a drift velocity equal to the linear flow velocity. These collisions and corresponding momentum exchanges will occur within one or, at most, a few mean free path lengths from the surface with the net effect of decreasing the linear velocity of those molecules flowing near the surface. As the area to volume ratio of the channels in the powder bed increases, the viscous drag effect will also increase and the rate of flow, for a given pressure gradient, will decrease. Clearly then, the surface area of the particles which constitute the channel walls is related to the gas flow rate and the pressure gradient. This type of flow is called viscous flow and is described by Poiseuille's law [37].

When the mean free path is approximately the same as the channel diameter, as with coarse powders at reduced gas pressures or fine powders

Permeametry

with the gas at atmospheric pressures, the gas will behave as though there were slippage at the channel walls. This occurs because collisions between molecules rebounding from a wall and the flowing molecules occur uniformly across the diameter of the channel. Therefore, there appears to be no preferential retardation of flow near the channel wall when compared to the center of the channel.

A third type of flow occurs at significantly reduced pressures where the mean free path of the gas molecules is greater than the channel diameter. Viscosity plays no part in this type of flow since the molecular collisions with the channel walls far outnumber the gas phase collisions. This type of molecular flow is a diffusion process.

Depending on which of the preceding three types of flow is employed, somewhat different measurements of the surface area of a sample is obtained. Viscous flow measurements tend to ignore blind pores and produce only the 'envelope' area of the particles. At the other extreme, diffusional flow senses the blind pores and often gives good agreement with surface areas measured by the BET method. At intermediate pressures, with so called slip flow, the very small blind pores are ignored while the larger ones can contribute to the measured surface area.

Darcy's law [38] asserts that the average flow velocity, v, of a fluid through a packed bed is proportional to the pressure gradient, ΔP, across the bed and is inversely proportional to the length, ℓ, of the bed. Thus,

$$v = K \frac{\Delta P}{\ell} \tag{7.13}$$

Poiseuille [37] has shown that the viscosity, η, of a fluid can be expressed in terms of the volume V (cm³) flowing through a tube of radius r (cm) and length ℓ (cm) in time t(s) under a pressure gradient of ΔP dyne cm^{-2}, that is,

$$\eta = \frac{\pi r^4 t \Delta P}{8 V \ell} \tag{7.14}$$

Equation (7.14) can be rearranged, by substituting $\pi r^2 \ell$ for V and v for ℓ/t

$$v = \frac{D^2 \Delta P}{32 \eta \ell} \tag{7.15}$$

where v is the linear flow velocity and D is the tube diameter. The similarity in form of equations (7.13) and (7.15) suggests that flow through a packed powder bed is equivalent to fluid through many small capillaries or channels. Kozeny [39] recognized this equivalency when he derived an equation for flow through a packed bed. Kozeny used the ratio of volume V to the area A of a tube, viz.,

48 Other surface area methods

$$\frac{V}{A} = \frac{D}{4} \tag{7.16}$$

If many parallel capillaries of equal length and diameters are used to carry the fluid, the ratio of V to A remains constant and equation (7.16) becomes

$$D_e = \frac{4V}{A} \tag{7.17}$$

where D_e is the diameter equivalent to one large tube.

In a powder bed, the void volume V_v is defined as the volume not occupied by solids. Further, the porosity p is defined by

$$P = \frac{V_v}{V_s + V_v} \tag{7.18}$$

where $V_s + V_v$ is the total volume of the powder bed and V_s is the volume occupied by the solid. Rearranging equation (7.18) gives

$$V_v = \left(\frac{p}{1-p}\right) V_s \tag{7.19}$$

Substituting V_v from equation (7.19) into equation (7.17) gives

$$D_e = 4\left(\frac{p}{1-p}\right)\frac{V_s}{A} \tag{7.20}$$

and replacing D in equation (7.15) with D_e yields

$$v_p = \left(\frac{V_s^2 \Delta P}{A^2 2\eta \ell_p}\right)\left(\frac{p}{1-p}\right)^2 \tag{7.21}$$

The term v_p is the average flow velocity through the channels in the bed and ℓ_p is the average channel length.

Because the approach velocity v, that is, the gas velocity prior to entering the powder bed is experimentally simple to measure, it is desirable to state equation (7.21) in terms of v rather than v_p. In order to make this conversion consider Fig. 7.2. The unshaded area represents gas contained between a piston and the surface of the powder bed. If the porosity p of the bed is 0.5, then in order to contain an equal volume of gas, the powder bed must be twice as long as the distance the piston can move. Clearly then, if the piston is displaced downward at a uniform rate the linear velocity through the bed must be twice the approach velocity, or

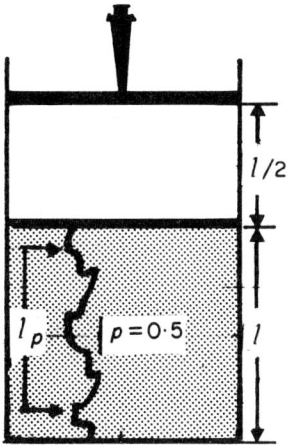

Figure 7.2 Piston displacement forcing gas to flow along a tortuous path through a packed powder bed.

$$v_p = \frac{v}{p} \tag{7.22}$$

However, equation (7.22) assumes that the channels in the powder bed are straight through. In fact, they are tortuous in shape, as indicated in the diagram. Actually, in the time required to travel through the powder bed an element of gas volume traverses a distance ℓ_p/ℓ greater than the length of the powder bed. This increases v_p and equation (7.22) is corrected to

$$v_p = \frac{v}{p}\left(\frac{\ell_p}{\ell}\right) \tag{7.23}$$

Replacing v_p in equation (7.21) with v from equation (7.23) gives

$$v = p\left(\frac{\ell}{\ell_p}\right)\left(\frac{V_s^2 \Delta P}{A^2 2\eta \ell_p}\right)\left[\frac{p^2}{(1-p)^2}\right] \tag{7.24}$$

and

$$A^2 = \frac{1}{2}\left(\frac{\ell}{\ell_p}\right)^2 \left(\frac{V_s^2 \Delta P}{\eta v \ell}\right)\left[\frac{p^3}{(1-p)^2}\right] \tag{7.25}$$

Further, if V_s^2 is replaced by the ratio of mass m to density ρ, equation (7.25) becomes

$$\left(\frac{A}{m}\right)^2 = \frac{1}{2}\left(\frac{\ell}{\ell_p}\right)^2 \left(\frac{\Delta P}{\rho^2 \eta \ell v}\right)\left[\frac{p^3}{(1-p)^2}\right] \tag{7.26}$$

The area A represents the total surface area of all the channels in the powder bed which offer resistance to flow. This excludes the surface area generated by close-ended or blind pores and crevices since they make no contribution to the frictional area. It follows then that equation (7.26) will give surface area values always less than BET or other adsorption techniques.

Recognizing that the ratio A/m in equation (7.26) is the specific surface area and calling $2(\ell_p/\ell)^2$ the 'aspect factor' f, which is determined by the particle's geometry, equation (7.26) can be written as

$$S^2 = \left(\frac{\Delta P}{\ell}\right)\left(\frac{p^3}{(1-p)^2}\right)\left(\frac{1}{f\rho^2 \eta v}\right) \tag{7.27}$$

Equation (7.27) will hold for incompressible fluids and for compressible fluids with small values of ΔP. If the pressure gradient across the bed is large and the fluid is compressible, equation (7.27) takes the form

$$S^2 = \left(\frac{P_2^2 - P_1^2}{2P_1}\right)\left(\frac{p^3}{(1-p)^2}\right)\left(\frac{1}{f\rho^2 \ell \eta v}\right) \tag{7.28}$$

where P_1 and P_2 are the exit and inlet pressures, respectively.

Experimentally the terms in equations (7.27) or (7.28) can be determined as follows:

1. ΔP can be measured by placing a sensitive pressure gauge or manometer at the inlet to the powder bed while venting to atmospheric pressure.
2. ρ, the true powder density, can be determined by a variety of methods, one of which is discussed in Chapter 22.
3. v can be determined by dividing the volumetric flow rate by the cross-sectional area of the powder bed.
4. p is calculated by subtracting the volume of powder from the total bed volume and dividing by the total bed volume.

The aspect factor, f, equal to $2(\ell_p/\ell)^2$ arises from the size and shape of the cross-sectional areas that make up the channels. Therefore, f is highly dependent upon the particle size and shape. Carman [40] studied numerous materials and found a value of 5 was suitable in most cases. Various theoretical and experimental values for 'f' usually are found to be in the range of 3 to 6.

Permeation measurements will be prone to serious errors if the average channel diameter is not within a factor of two of the largest diameter.

Large diameter channels tend to give area to volume ratios excessively low (compare equation (7.17)), while their contribution to the average flow rate is in excess of their number. Also, agglomerated particles behave as one particle, the flow measuring only the envelope of the aggregate.

Several modifications of Poiseuille's equation have been attempted by various authors [41–43] to describe permeability in the transitional region between viscous and diffusional flow. The assumptions underlying these modifications are often questionable and the results obtained offer little or no theoretical or experimental advantage over the BET theory for surface area measurements. Allen [44] discusses these modifications as well as diffusional flow at low pressures.

Viscous flow permeametry measured near atmospheric pressure offers the advantages of experimental simplicity and a means of measuring the external or envelope area of a powder sample which is otherwise not readily available by any adsorption method. The usefulness of measuring the external surface area rather than the BET or total surface area becomes evident if the data is to be correlated with fluid flow through a powder bed or with the average particle size.

8
Pore analysis by adsorption

8.1 THE KELVIN EQUATION

Adsorption studies leading to measurements of pore size and pore size distributions generally make use of the Kelvin equation [45] which relates the equilibrium vapor pressure of a curved surface, such as that of a liquid in a capillary or pore, to the equilibrium pressure of the same liquid on a plane surface. Equation (8.1) is a convenient form of the Kelvin equation

$$\ln \frac{P}{P_0} = -\frac{2\gamma \bar{V}}{rRT} \cos \theta \tag{8.1}$$

where P is the equilibrium vapor pressure of the liquid contained in a narrow pore of radius r and P_0 is the equilibrium pressure of the same liquid exhibiting a plane surface. The terms γ and $\bar{V}$ are the surface tension and molar volume of the liquid, respectively, and θ is the contact angle with which the liquid meets the pore wall.

In a pore, the overlapping potentials of the walls more readily overcome the translational energy of an adsorbate molecule so that condensation will occur at a lower pressure in a pore than that normally required on an open or plane surface. Thus, as the relative pressure is increased, condensation will occur first in pores of smaller radii and will progress into the larger pores until, at a relative pressure of unity, condensation will occur on those surfaces where the radius of curvature is essentially infinite. Conversely, as the relative pressure is decreased, evaporation will occur progressively out of pores with decreasing radii.

In order to derive the Kelvin equation on thermodynamic grounds, consider the transfer of dn moles of vapor in equilibrium with the bulk liquid at pressure P_0 into a pore where the equilibrium pressure is P. This process consists of three steps: evaporation from the bulk liquid, expansion of the vapor from P_0 to P and condensation into the pore. The first and third of these steps are equilibrium processes and are therefore accompanied by a zero free energy change, whereas the free energy change for the second step is described by

$$dG = \left(RT \ln \frac{P}{P_0}\right) dn \tag{8.2}$$

When the adsorbate condenses in the pore it does so on a previously adsorbed film thereby decreasing the film-vapor interfacial area. The free energy change associated with the filling of the pore is given by

$$dG = -(\gamma \cos \theta) \, dS \tag{8.3}$$

where γ is the surface tension of the adsorbed film, assumed to be identical with that of the liquid, dS is the change in interfacial area and θ is the wetting angle which is taken to be zero since the liquid is assumed to wet completely the adsorbed film. Fig. 8.1 illustrates the difference between a zero and a finite wetting angle.

Equations (8.2) and (8.3), when combined using the assumption of a zero wetting angle, yield

$$\frac{dn}{dS} = \frac{-\gamma}{RT \ln P/P_0} \tag{8.4}$$

The volume of liquid adsorbate which condenses in a pore of volume V_p is given by

$$dV_p = \bar{V} dn \tag{8.5}$$

where $\bar{V}$ is the molar volume of the liquid adsorbate. Substituting equation (8.5) into equation (8.4) gives

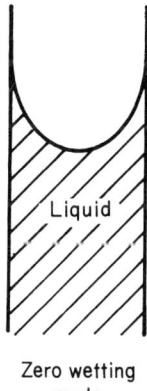

Zero wetting angle

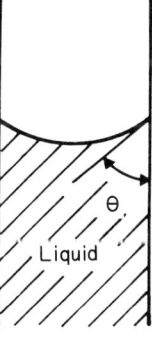

Finite wetting angle

Figure 8.1 Wetting angles.

$$\frac{dV_p}{dS} = \frac{-\gamma \bar{V}}{RT \ln P/P_0} \tag{8.6}$$

The ratio of volume to area within a pore depends upon the pore geometry. For example, the volume to area ratios for cylinders, parallel plates and spheres are, respectively, $r/2$, $r/2$ and $r/3$, where r is the cylinder and sphere radii or the distance of separation between parallel plates. If the pore shapes are highly irregular or consist of a mixture of regular geometries, the volume to area ratio can be too complex to express mathematically. In these cases, or in the absence of specific knowledge of the pore geometry, the assumption of cylindrical pores is usually made, and equation (8.6) becomes

$$\ln \frac{P}{P_0} = \frac{-2\gamma \bar{V}}{rRT} \tag{8.7}$$

Equation (8.7) is the working equation for pore size analysis by adsorption unless more specific information is available regarding pore geometry and the wetting angle.

Of the five isotherm classifications depicted in Fig. 3.1 the Types I, IV and V isotherms are associated with porosity. The Type I isotherm usually corresponds to microporosity, that is, pores of diameters only slightly larger than adsorbate molecules. The Types IV and V isotherms are associated with pores ranging in radius from about fifteen to several hundred ångströms. The Type IV isotherm is more frequently encountered with porous adsorbents.

The Types IV and Type V isotherms resemble the Type II and Type III isotherms, respectively, at the bottom or low pressure end. However, at the high pressure end they turn toward the $P/P_0 = 1$ line and as indicated in Fig. 8.2 the isotherm can approach the saturated pressure axis asymptotically (D–G) or at a finite angle (D–E).

In those cases where the isotherm cuts the saturated pressure line at a finite angle, the liquid condensed in the pores does not completely wet the previously adsorbed film. This condition is presumed to arise because the film persists in exhibiting a different structure than the liquid due to the extended influence of the adsorbent's surface potential. In this case it would be more correct to use equation (8.1) which includes the wetting angle, rather than equation (8.7) which assumes a zero wetting angle. However, for wetting angles even as large as 20° the value of $\cos \theta$ is 0.94. Thus, the error caused by assuming a zero wetting angle will generally be small, particularly when compared to the errors introduced by the lack of knowledge regarding the exact pore shape.

When the isotherm approaches the saturated pressure line asymptotically, which is the prevalent condition, wetting of the previously adsorbed film by the adsorbate condensed in the pores is assured and the wetting angle may be taken with certainty as zero.

8.2 ADSORPTION HYSTERESIS

Fig. 8.2 also shows hysteresis which is typical of type IV isotherms. The curve BCD indicates the path traversed along the adsorption isotherm as the relative pressure is increased, while the curve DFB shows the path followed along the desorption isotherm as the relative pressure is reduced. The presence of the hysteresis loop introduces a considerable complication, in that within the region of the hysteresis loop there are two relative pressure values corresponding to a given quantity adsorbed, with the lower value always residing on the desorption isotherm. As shown in Fig. 8.2, the quantity W is adsorbed at a lower relative pressure on the desorption isotherm than on the adsorption curve. The pore radius corresponding to the quantity adsorbed must be single-valued and some criteria have to be established as to which value of relative pressure should be employed in the Kelvin equation. As shown previously (cf. equation (8.2)), the molar free energy change accompanying the condensation of vapor into a pore during adsorption is given by

$$\Delta G_{ads} = RT(\ln P_{ads} - \ln P_0) \tag{8.8}$$

For the same quantity of adsorbate on the desorption isotherm the corresponding free energy change is

$$\Delta G_{des} = RT(\ln P_{des} - \ln P_0) \tag{8.9}$$

Since $P_{des} < P_{ads}$ it follows that $\Delta G_{des} < \Delta G_{ads}$. Therefore, the desorption value of relative pressure corresponds to the more stable adsorbate

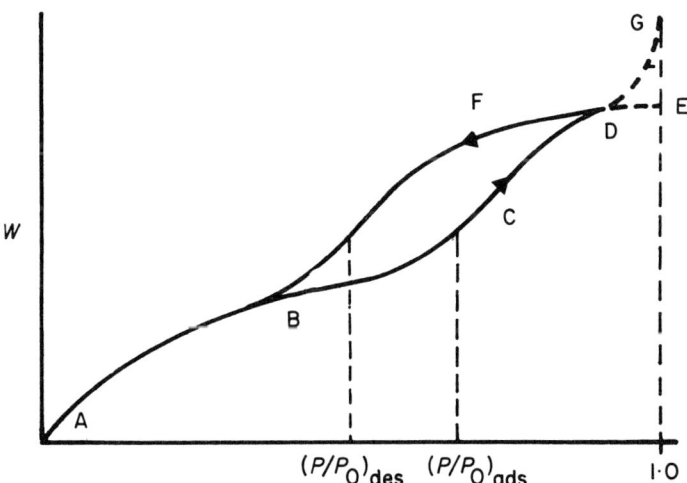

Figure 8.2 Typical Type IV adsorption and desorption isotherms showing hysteresis.

56 Pore analysis by adsorption

condition and the desorption isotherm should, with certain exceptions, be used for pore size analysis.

Several theories have been formulated in order to explain the difference between the state of the adsorbate during adsorption and that during desorption. For example, Zsigmondy [46] postulated that hysteresis was caused by a difference of contact angle during adsorption and desorption.

McBain [47] accounted for hysteresis by assuming that the pores contained a narrow opening and a wide body, the so-called 'bottle-neck' shape. His model asserts that during adsorption the wide inner portion of the pore is filled at high relative pressures but cannot empty until the narrow neck of the pore first empties at lower relative pressures during desorption. Therefore, for 'bottle-neck' pores the adsorption isotherm corresponds to the equilibrium condition. However, the model proposed by McBain ignores the question of how condensation into the wider inner portion of the pore can occur once the narrow neck has been filled at low relative pressure.

Cohan [48] assumed that condensation occurs by filling the pore from the wall inward which, for a cylindrical pore, would give a cylindrically shaped meniscus, whereas evaporation occurs from a hemispherical meniscus once the pore is filled.

Foster [49] explained hysteresis by considering that the pores fill by adsorption on the walls, while emptying by evaporation from a spherical meniscus.

These and other theories [50] may each describe a unique condition leading to hysteresis and indeed there may be no single mechanism which can universally explain the phenomenon.

8.3 TYPES OF HYSTERESIS

Rarely have hysteresis loops been found which have not closed, at the low end, usually by relative pressures of 0.3. According to the Kelvin equation, pore radii corresponding to relative pressures less than 0.3 would be smaller than 15 Å. Since monolayer formation is usually complete when the relative pressure reaches 0.3, the radius available for condensation would be diminished by the thickness of the monolayer or by about two molecular diameters. The radius of this center core would then be approximately one or two molecular diameters. It would be difficult to establish the validity of the Kelvin equation for pores this small, and it is entirely possible that adsorption and desorption in micropores proceed by the same path, thereby precluding hysteresis.

At relative pressures above 0.3, de Boer [51] has identified five types of hysteresis loops which he correlated with various pore shapes. Fig. 8.3 shows idealizations of the five types of hysteresis.

Type A hysteresis is due principally to cylindrical pores open at both ends. According to Cohan [48], type A hysteresis is caused by condensation pro-

Types of hysteresis 57

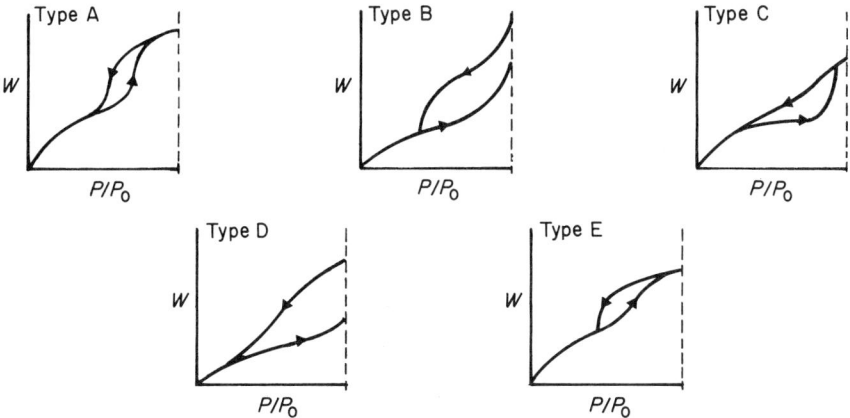

Figure 8.3 de Boer's five types of hysteresis.

ducing a cylindrical meniscus with one radius of curvature equal to the pore radius, less the thickness of previously condensed film, and the other radius is the length of the pore or essentially infinite. During desorption, Cohan views the meniscus as being hemispherical. According to this model, as the cylindrical film increases in depth during adsorption, the changes in the remaining volume and area are given, respectively, by

$$dV = -2\pi r_k \ell dr_k \tag{8.10}$$

and

$$dS = -2\pi \ell dr_k \tag{8.11}$$

where r_k is the radius of the unfilled part, or center core, of the cylindrical pore and ℓ is the pore length. Combining equations (8.10) and (8.11) gives

$$\frac{dV}{dS} = r_k \tag{8.12}$$

Placing the above value for dV/dS into equation (8.6) yields a modified form of the Kelvin equation, namely,

$$\ln\left(\frac{P}{P_0}\right)_{ads} = -\frac{\gamma \overline{V}}{r_k RT} \tag{8.13}$$

During desorption the center core evaporates from a hemispherical meniscus; therefore, the Kelvin equation is applicable.

58 Pore analysis by adsorption

$$\ln\left(\frac{P}{P_0}\right)_{des} = -\frac{2\gamma\bar{V}}{r_k RT} \qquad \text{(cf.8.7)}$$

From equations (8.13) and (8.7), it is evident that condensation into and evaporation out of the inner core will occur when

$$\left(\frac{P}{P_0}\right)^2_{ads} = \left(\frac{P}{P_0}\right)_{des} \qquad (8.14)$$

Therefore, for a real system of pores of approximately cylindrical geometry it is expected that pores of a given radius will fill at a higher relative pressure than they will empty.

The type B hysteresis curve is associated with slit-shaped pores or the space between parallel plates.

Type C hysteresis is produced by a mixture of tapered or wedge-shaped pores with open ends.

Type D curves are also produced by tapered or wedge-shaped pores but with narrow necks at one or both open ends.

Type E hysteresis results from McBain's 'bottle-neck' pores. In pores of this shape, emptying of the wide portion will be delayed during desorption until the narrow neck can evaporate. Therefore, the desorption curve exhibits a small slope at high relative pressures and a large slope where the wide part of the pore empties.

8.4 TOTAL PORE VOLUME

If W_s is the adsorbate weight at saturation (i.e., when $P/P_0 = 1$), then $W_s/\rho_\ell = V_\ell$, where V_ℓ and ρ_ℓ are the volume of liquid adsorbate at saturation and the liquid density, respectively. Various studies [52–55] have shown that at saturation the liquid volume of different adsorbates, when measured on porous adsorbents, is essentially constant and is independent of the adsorbate. This constancy of adsorbed liquid at saturation is known as the Gurvitsh [56] rule and provides direct evidence that the pores are filled with liquid adsorbate at saturated vapor pressures.

To calculate the total pore volume it is necessary to measure the total adsorbate volume at a relative pressure above the point where the hysteresis loop closes. In general, a relative pressure as close to unity as possible should be chosen in order to include the large radii pores in the measurement. For example, if W_a grams of nitrogen are adsorbed at $P/P_0 = 0.99$ then the corresponding volume of pores V_p and the largest pore radii r are given by

$$V_p = \frac{W_a}{\rho_\ell} \qquad (8.15)$$

and from the Kelvin equation

$$r = \frac{-2(8.85)(34.6)}{(8.314 \times 10^7)(77)(2.303)\log 0.99} = 950 \times 10^{-8}\,\text{cm} \quad (8.16)$$

where $8.85\,\text{erg}\,\text{cm}^{-2}$ is the surface tension and $34.6\,\text{cm}^3$ is the molar volume of liquid nitrogen at 77 K. Equations (8.15) and (8.16) state that the total volume of all the pores up to 950 Å is $V_\text{p}\,\text{cm}^3$. An implicit assumption in these calculations is that no surface other than the inner walls of the pores exists. Using this assumption and that of cylindrical geometry, the average pore radius $\bar{r}_\text{p}$ can be calculated from the ratio of the total pore volume and the BET surface area from the following equation:

$$\frac{V_\text{p}}{S_\text{BET}} = \frac{\bar{r}_\text{p}}{2} \quad (8.17)$$

8.5 PORE SIZE DISTRIBUTIONS

On thermodynamic grounds pore size distributions are measured using the desorption isotherm (see equations (8.8) and (8.9)). The exceptions to this are 'bottle-neck' pores exhibiting type E hysteresis. In this case, the equilibrium isotherm is that of adsorption because the unstable state is associated with the condition that the wide portion of the pore is unable to evaporate until the narrow neck empties. Regardless, however, of which isotherm is used the mathematical treatment remains the same.

For nitrogen as the adsorbate at its normal boiling point of 77 K, the Kelvin equation can be written as

$$r_\text{k} = \frac{4.15}{\log P_0/P}(\text{Å}) \quad (8.18)$$

using the physical properties given in equation (8.16). The term r_k indicates the radius into which condensation occurs at the required relative pressure. This radius, called the Kelvin radius or the critical radius, is not the actual pore radius since some adsorption has already occurred on the pore wall prior to condensation, leaving a center core or radius r_k. Alternatively, during desorption, an adsorbed film remains on the pore wall when evaporation of the center core takes place.

If the depth of the adsorbed film when condensation or evaporation occurs is t, then the actual pore radius r_p is given by

$$r_\text{p} = r_\text{k} + t \quad (8.19)$$

Equation (8.18) can be used to calculate r_k but some means of evaluating t is required if the pore radius is to be determined.

Using the assumption that the adsorbed film depth in a pore is the same as that on a plane surface for any value of relative pressure, one can write

$$t = \left(\frac{W_a}{W_m}\right)\tau \qquad (8.20)$$

where W_a and W_m are, respectively, the quantity adsorbed at a particular relative pressure and the weight corresponding to the BET monolayer. Essentially, equation (8.20) asserts that the thickness of the adsorbed film is simply the number of layers times the thickness τ of one layer regardless of whether the film is in a pore or on a plane surface.

The value of τ can be calculated by considering the area S and volume $\bar{V}$ occupied by one mole of liquid nitrogen if it were spread over a surface to the depth of one molecular layer.

$$S = (16.2)(6.02 \times 10^{23}) = 97.5 \times 10^{23}\,\text{Å}^2 \qquad (8.21)$$

$$\bar{V} = (34.6 \times 10^{24})\,\text{Å}^3 \qquad (8.22)$$

Then

$$\tau = \frac{\bar{V}}{S} = 3.54\,\text{Å} \qquad (8.23)$$

This value of 3.54 Å is somewhat less than the diameter of a nitrogen molecule based upon the cross-sectional area of 16.2 Å^2. This is as it should be since the liquid structure is considered to be closed packed hexagonal and each nitrogen molecule sits in the depression between three molecules in the layers above and below. Equation (8.20) can now be written as

$$t = \left(\frac{W_a}{W_m}\right) 3.54\,\text{Å} \qquad (8.24)$$

On nonporous surfaces it has been shown that when W_a/W_m is plotted versus P/P_0 the data all approximately fit a common type II curve above a relative pressure of 0.3 [57–62]. This implies that when $W_a/W_m = 3$, for example, the adsorbed layer thickness t will be 10.62 Å regardless of the adsorbent. The common curve is described closely by the Halsey [63] equation which for nitrogen can be written as

$$t = 3.54 \left(\frac{5}{2.303 \log P_0/P}\right)^{\frac{1}{3}} \tag{8.25}$$

To calculate the pore size distribution consider the work sheet shown as Table 8.1, and the corresponding explanation of each column. The adsorbed volumes are from a hypothetical isotherm. The procedure used is the numerical integration method of Pierce [60] as modified by Orr and Dalla Valle [64] with regard to calculating the depth of the adsorbed film. This method as well as the Barrett, Joyner and Halenda (BJH) [65] numerical integration method takes advantage of Wheeler's [66] theory that condensation occurs in pores when a critical relative pressure is reached corresponding to the Kelvin radius r_k. This model also assumes that a multilayer of adsorbed film of depth t exists on the pore wall when evaporation or condensation occurs which is of the same depth as the adsorbed film on a nonporous surface.

The method shown in Table 8.1 uses data from either the adsorption or desorption isotherm. However, as stated previously, the desorption curve is usually employed except in those cases where the adsorption curve corresponds to the thermodynamically more stable condition such as in the case of 'bottle-neck' pores. In either case, for ease of presentation, the data is evaluated downward from high to low relative pressures.

Columns 1 and 2 of Table 8.1 contain data obtained directly from the isotherm. The desorbed volumes are normalized for one gram of adsorbent. Relative pressures are chosen using small decrements at high values, where r_k is very sensitive to small changes in relative pressure and where the slope of the isotherm is large, such that small changes in relative pressure produce large changes in volume.

Column 3, the Kelvin radius, is calculated from the Kelvin equation assuming a zero wetting angle. If nitrogen is the adsorbate, equation (8.18) can be used.

Column 4, the film depth, t, is calculated using equation (8.25), the Halsey equation.

Column 5 gives the pore radius r_p obtained from equation (8.19).

Columns 6 and 7, $\bar{r}_k$ and $\bar{r}_p$, are prepared by calculating the mean value in each decrement from successive entries.

Column 8, the change in film depth is calculated by taking the difference between successive values of t.

Column 9, ΔV_{gas}, is the change in adsorbed volume between successive P/P_0 values and is determined by subtracting successive values from column 2.

Column 10, ΔV_{liq}, is the volume of liquid corresponding to ΔV_{gas}. The most direct way to convert ΔV_{gas} to ΔV_{liq} is to calculate the moles of gas, and multiply by the liquid molar volume. For nitrogen at standard temperature and pressure, this is given by

Table 8.1 Pore size distribution work table

1	2	3	4	5	6	7	8	9	10	11	12	13	14
P/P_0	V_{STP}^{gas} (cm³ g⁻¹)	r_k (Å)	t (Å)	r_p (Å)	$\bar{r}_k$ (Å)	$\bar{r}_p$ (Å)	Δt (Å)	ΔV_{STP}^{gas} (cm³ g⁻¹)	$\Delta V_{liq} \times 10^3$ (cm³ g⁻¹)	$\Delta t \Sigma S \times 10^3$ (cm³ g⁻¹)	$V_p \times 10^3$ (cm³ g⁻¹)†	S (m²)	ΣS (m²)‡
0.99	161.7	950	28.0	978									
					711	737	5.8	0.2	0.31	0	0.33	0.01	0.01
0.98	161.5	473	22.2	495									
					394	414	2.8	0.5	0.77	0.00	0.85	0.04	0.05
0.97	161.0	314	19.4	333									
					250	268	3.1	0.8	1.23	0.02	1.40	0.10	0.15
0.95	160.2	186	16.3	202									
					138	153	3.5	1.4	2.16	0.05	2.59	0.34	0.49
0.90	158.8	90.7	12.8	104									
					74.8	87.0	1.7	1.6	2.46	0.08	3.22	0.74	1.23
0.85	157.2	58.8	11.1	69.9									
					50.8	61.4	1.1	2.0	3.08	0.14	4.30	1.40	2.63
0.80	155.2	42.8	10.0	52.8									
					39.7	49.5	.5	2.3	3.54	0.13	5.30	2.14	4.77
0.77	152.9	36.6	9.5	46.1									
					34.9	44.3	.3	4.0	6.16	0.14	9.70	4.38	9.15
0.75	148.9	33.2	9.2	42.4									
					31.8	40.9	.3	3.8	5.85	0.27	9.22	4.51	13.66
0.73	145.1	30.4	8.9	39.3									
					29.2	38.0	.2	4.2	6.47	0.27	10.49	5.52	19.18
0.71	140.9	27.9	8.7	36.6									
					26.9	35.4	.3	5.0	7.70	0.58	12.34	6.97	26.15

0.69	135.9	25.8	8.4	34.2									
0.67	130.0	23.9	8.2	32.1	24.9	33.2	.2	5.9	9.09	0.52	15.23	9.17	35.32
0.65	123.9	22.2	8.0	30.2	23.1	31.2	.2	6.1	9.39	0.71	15.84	10.15	45.47
0.63	117.3	20.7	7.8	28.5	21.5	29.4	.2	6.6	10.16	0.91	17.30	11.77	57.24
0.61	110.1	19.3	7.7	27.0	20.0	27.8	.1	7.2	11.09	0.57	20.32	14.62	71.86
0.59	102.6	18.1	7.5	25.6	18.7	26.3	.2	7.5	11.55	1.44	20.00	15.21	87.07
0.57	95.0	17.0	7.3	24.3	17.6	25.0	.2	7.6	11.70	1.74	20.09	16.07	103.1
0.55	86.9	16.0	7.2	23.2	16.5	23.8	.1	8.1	12.47	1.03	23.80	20.00	123.1
0.53	78.8	15.1	7.0	22.1	15.6	22.7	.2	8.1	12.47	2.46	21.19	18.67	141.8
0.51	71.5	14.2	6.9	21.1	14.7	21.6	.1	7.3	11.24	1.42	21.21	19.64	161.4
0.49	65.4	13.4	6.8	20.2	13.8	20.7	.1	6.1	9.39	1.61	17.50	16.90	178.3
0.45	57.3	12.0	6.5	18.5	12.7	19.4	.3	8.1	12.47	5.35	16.62	17.13	195.4
0.40	51.7	10.4	6.2	16.6	11.2	17.6	.3	5.6	8.62	5.86	6.81	7.74	203.1
0.35	47.4	9.1	6.0	15.1	9.8	15.9	.2	4.3	6.62	4.06	6.73	8.47	212.1

† $\Sigma V_P = 0.28 \, \text{cm}^3 \, \text{g}^{-1}$; ‡ $\Sigma S = 212.1 \, \text{m}^2 \, \text{g}^{-1}$

$$\Delta V_{\text{liq}} = \frac{\Delta V_{\text{gas}}}{22.4 \times 10^3} \times 34.6 = \Delta V_{\text{gas}}(1.54 \times 10^{-3})\,\text{cm}^3 \tag{8.26}$$

Column 11 represents the volume change of the adsorbed film remaining on the walls of the pores from which the center core has previously evaporated. This volume is the product of the film area ΣS and the decrease in the film depth Δt. By assuming no pores are present larger than 950 Å ($P/P_0 = 0.99$) the first entry in column 11 is zero since there exists no film area from previously emptied pores. The error introduced by this assumption is negligible because the area produced by pores larger than 950 Å will be small compared to their volume.

Subsequent entries in column 11 are calculated as the product of Δt for a decrement and ΣS from the row above corresponding to the adsorbed film area exposed by evaporation of the center cores during all the previous decrements.

Column 12, the actual pore volume, is evaluated by recalling that the volume of liquid, column 10, is composed of the volume evaporated out of the center cores plus the volume desorbed from the film left on the pore walls. Then,

$$\Delta V_{\text{liq}} = \pi \bar{r}_k^2 \ell + \Delta t \Sigma S \tag{8.27}$$

and since,

$$V_p = \pi \bar{r}_p^2 \ell \tag{8.28}$$

where ℓ is the pore length, by combining the above two equations,

$$V_p = \left(\frac{\bar{r}_p}{\bar{r}_k}\right)^2 [\Delta V_{\text{liq}} - (\Delta t \Sigma S)(10^{-4})]\,\text{cm}^3 \tag{8.29}$$

Column 13 is the surface area of the pore walls calculated from the pore volume by

$$S = \frac{2V_p}{\bar{r}_p} \times 10^4\,(\text{m}^2) \tag{8.30}$$

with V_p in cubic centimeters and $\bar{r}_p$ in ångströms. It is this value of S which is summed in column 14. The summation is multiplied by Δt from the following decrement to calculate the film volume decrease in column 11.

If the utmost rigor were used, it would be correct to modify the area contributed by previously emptied pores since their statistical thickness diminishes with each successive decrement. However, this procedure would be cumbersome and of questionable value in view of the many other assumptions which have been made. Nevertheless, the BJH [65] method attempts to make this modification by introducing an average inner core based on its variation with each decrement of relative pressure.

Table 8.1 discloses that the volume of all pores greater than 15.1 Å is 0.28 cm^3 g^{-1}. This does not mean that micropores with smaller radii are absent but rather, as stated earlier, that the validity of the Kelvin equation becomes questionable because of the uncertainty regarding molar volumes and surface tension when only one or two molecular diameters are involved [67–68]. Termination of the analysis is necessary, therefore, when the relative pressure approaches 0.3. In some analyses, it is found that the volume desorbed from the film ($\Delta t \Sigma S$) becomes equal to the total amount desorbed, indicating that only desorption is occurring in the narrower pores and the analysis must be ended because in the absence of evaporation from center cores the Kelvin equation is not applicable. A third condition necessitating termination of the analysis is encountered when the hysteresis loop closes at higher relative pressures indicating the absence of pores below that relative pressure.

The volume of micropores, if present, can be evaluated by the difference between the total pore volume (see equation (8.15)) and the sum of column 12, ΣV_p.

The total area of all pores to 15.1 Å radius (column 14) is 212.1 m^2 g^{-1}. This area is usually less than the BET area since it does not include the surface contributed by micropores. An area larger than the BET area would be exhibited by ink-bottle pores in which a larger volume of gas is condensed in pores having a relatively small area. This is the case shown in the example in Table 8.1.

An implicit assumption, hidden in the method, is that no surface exists other than that within pores because the volume desorbed in any decrement is assumed to originate from either the center core or the adsorbed film on the pore wall.

Fig. 8.4 is the pore size distribution curve prepared from the data contained in Table 8.1. The terms ΔV_p and Δr_p are taken from columns 12 and 5, respectively. Values of $\bar{r}_p$ are from column 7. Fig. 8.5 is the same data plotted as the cumulative pore volume.

8.6 MODELLESS PORE SIZE ANALYSIS

The ideas of Wheeler [66] that condensation and evaporation occur within a center pore during adsorption and desorption and that an adsorbed film is present on the pore wall has led to the proposal of various methods for pore size analysis. In addition to the methods of Pierce [60] and the BJH [65] technique, other schemes have been proposed, including those by Shull [58].

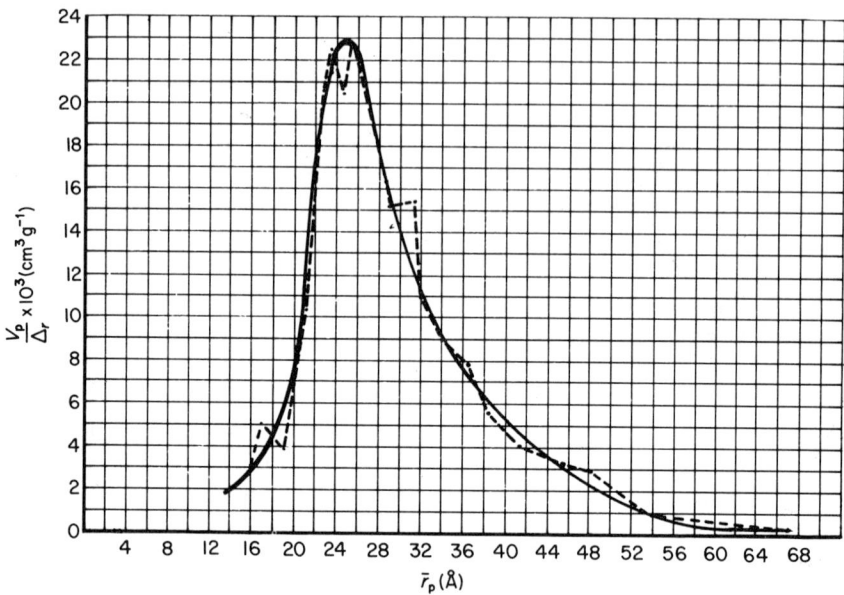

Figure 8.4 Pore size distribution curve from Table 8.1. Raw data: ----; smoothed data: ——.

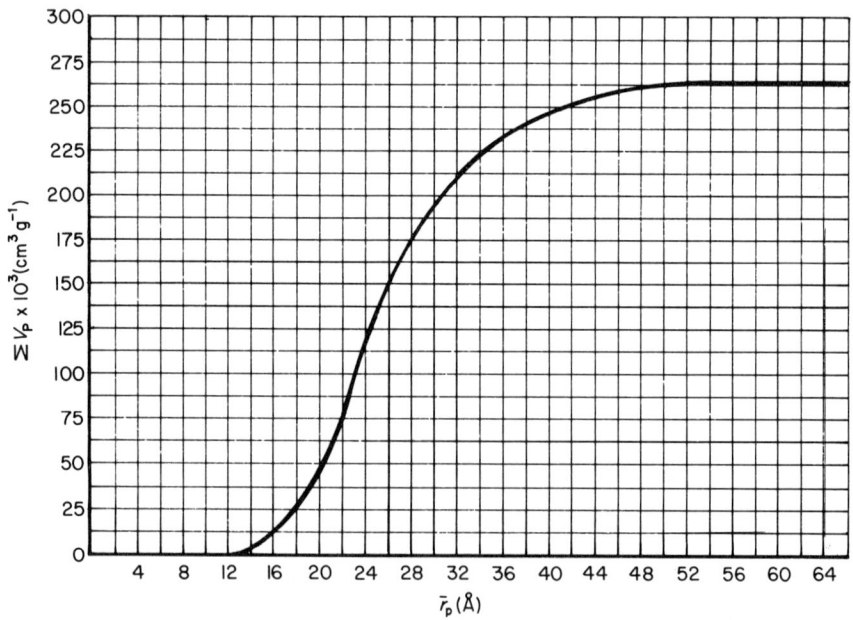

Figure 8.5 Cumulative pore volume plot from Table 8.1.

Oulton [69], Roberts [70], Innes [71], and Cranston and Inkley [62]. These ideas are all based upon some assumption regarding the pore shape.

Brunauer et al. [72], have developed a means of determining the pore volume distribution wherein the pore shape has a negligible influence.

Equation (8.4) establishes the relationship between the moles of adsorbate condensed into pores and the corresponding decrease in the pore area. Rewriting equation (8.4) and eliminating the negative sign, since during condensation pore surface is disappearing, one obtains

$$S = \frac{1}{\gamma} \int \left(RT \ln \frac{P}{P_0}\right) dn \qquad (8.31)$$

Kiselev [73], using the above equation by graphical integration of the isotherm between the limits of saturation and hysteresis loop closure, was able to calculate surface areas for wide pore samples in good agreement with BET measured areas. For micropores, the absence of hysteresis at the low pressure end of the isotherm indicates that only adsorption and not condensation occurs, thereby rendering Kiselev's method inapplicable.

Brunauer's modelless method uses pore volume and pore area not as functions of the Kelvin radius but rather as functions of hydraulic radii that he defines as

$$r_h = \frac{V}{S} \qquad (8.32)$$

where V and S are the pore volume and surface, respectively, regardless of their shape. For cylinders and parallel plates, r_h is one-half of radius or distance between plates.

As in the method of Pierce [60], Brunauer assumes that in the first decrement, say between $P/P_0 = 1$ and 0.95, desorption occurs only from the center core of the largest pores regardless of their shape. The hydraulic radius of the core is calculated by dividing the liquid volume which has evaporated out of the pore by the core area as determined by graphically integrating equation (8.31) using $P/P_0 = 1$ and 0.95 as the limits of integration.

In the second decrement the liquid volume desorbed $(V_{\text{liq}})_2$ must be corrected for the decrease in the adsorbed film depth remaining on the walls of previously emptied pores. By assuming the pores are cylindrical, the core volume $(V_k)_2$ can be calculated from the decrease in statistical thickness t, as

$$(V_k)_2 = (V_{\text{liq}})_2 - S_1(t_1 - t_2) - \frac{S_1}{4(r_h)_1}(t_1 - t_2)^2 \qquad (8.33)$$

where S_1 and $(r_h)_1$ are the core surface and hydraulic radius calculated from the first decrement. The terms t_1 and t_2 refer to the adsorbed film depths at the

68   Pore analysis by adsorption

beginning and end of the second decrement. These values can be obtained from equation (8.25), the Halsey equation. The volume $(V_k)_2$ is divided by S_2 calculated from equation (8.31) using as limits for the graphical integration the relative pressures at the beginning and end of the second decrement. This ratio gives the second hydraulic radius $(r_h)_2$.

To calculate the hydraulic radius in the third decrement the desorbed volume must be corrected for the contribution made by the film on the walls of pores emptied in the first decrement and decreased in depth during the second decrement plus the contribution made by the change in film depth within pores emptied in the second decrement. The total correction factor is

$$(V_k)_3 = (V_{liq})_3 - S_1(t_1 - t_3) - \frac{S_1}{4(r_h)_1}(t_1 - t_3)^2 - S_1(t_1 - t_2) - \frac{S_1}{4(r_h)_1}(t_1 - t_2)^2 - S_2(t_2 - t_3) - \frac{S_2}{4(r_h)_2}(t_2 - t_3)^2 \qquad (8.34)$$

It becomes painfully and quickly obvious that for an analysis requiring many data points, the number of correction terms becomes cumbersome. However, Brunauer was able to show that the squared terms do not contribute significantly and the above equation can be rewritten as

$$(V_k)_3 = (V_{liq})_3 - S_1(t_1 - t_3) - S_1(t_1 - t_2) - S_2(t_2 - t_3) \qquad (8.35)$$

Equation (8.35) is identical to the correction factor for parallel plate pores.

Using the above corrections, the Brunauer method is not modelless. However, it does offer a means of employing the same correction factor for pores as diverse as parallel plates and cylinders. In those instances where the pore geometry does differ considerably from cylinders or parallel plates, the error introduced by assuming either of these shapes is small. This was confirmed by Brunauer by plotting pore distributions, (V_k/r_h) versus r_h, using corrected and uncorrected core volumes. These plots differed only slightly and the hydraulic radius remained essentially unchanged. Accordingly, if uncorrected hydraulic core volumes are used the method is entirely modelless and little accuracy is sacrificed.

8.7   V–t CURVES

In Chapter 4, it was emphasized that the surface of an adsorbent is never covered with an adsorbed film of uniform depth but rather with stacks of adsorbed molecules of various depths. Discussions in this chapter regarding film depth t on the pore walls implies that the film is uniform. To be consistent with the BET theory, the adsorbed film on the wall of a pore is structured as on a nonporous surface and the film depth t is in reality a 'statistical depth' which is proportional to the number of monolayers present in the film regardless of how the molecules may be stacked.

Shull [58] showed that on a number of nonporous solids, the ratio of the weight adsorbed W_a and the weight corresponding to formation of a monolayer W_m when plotted versus the relative pressure could be closely represented by a single curve regardless of the solid. The curve produced by Shull is a typical Type II isotherm. Similar plots were made by others [57–62] on a variety of nonporous materials and in each case the points reasonably well fit the same common curve, particularly at relative pressures above 0.3. The common curve is also fit by the Halsey [63] equation above 0.3 relative pressure. If the monolayer is envisioned as being uniformly one molecule in depth, then a plot of W_a/W_m versus P/P_0 discloses the relative pressures corresponding to surface coverage by any number of monolayers. Therefore, if the adsorbate diameter is known, the statistical depth t can be calculated by multiplying the number of monolayers by the adsorbate diameter. Fig. 8.6 is a t plot of the statistical thickness versus the relative pressure, P/P_0, prepared from the Halsey equation. Experimental plots lie slightly above and below the indicated curve.

Shull [58] assumed that the adsorbate molecules were packed one on top of the other in the film and deduced the monolayer depth to be 4.3 Å for nitrogen. A more realistic assumption is that the film structure is close packed hexagonal, leading to a monolayer depth of 3.54 Å as shown by equation (8.23). Therefore, as stated previously, the statistical depth t of the adsorbed film is

$$t = \left(\frac{W_a}{W_m}\right) 3.54 \text{ Å} \tag{cf.8.24}$$

If the volume adsorbed on a surface of area S is expressed as the corresponding liquid volume, then

$$t = \frac{V_{liq}}{S} \times 10^4 \text{ (Å)} \tag{8.36}$$

with V_{liq} in cubic centimeters and S in square meters.

Lippens and de Boer [59] have shown that a plot of the volume adsorbed V_{liq} versus t calculated from equation (8.24) will yield a straight line of slope $S \times 10^{-4}$ from which the surface area can be calculated. Plots of this nature are termed V–t curves and generally give surface areas comparable to BET values provided that multilayer adsorption and not condensation into the pores takes place.

Figure 8.7 illustrates three types of V–t curves. Curve X results from multilayer adsorption in the absence of any condensation into pores. Curve Y

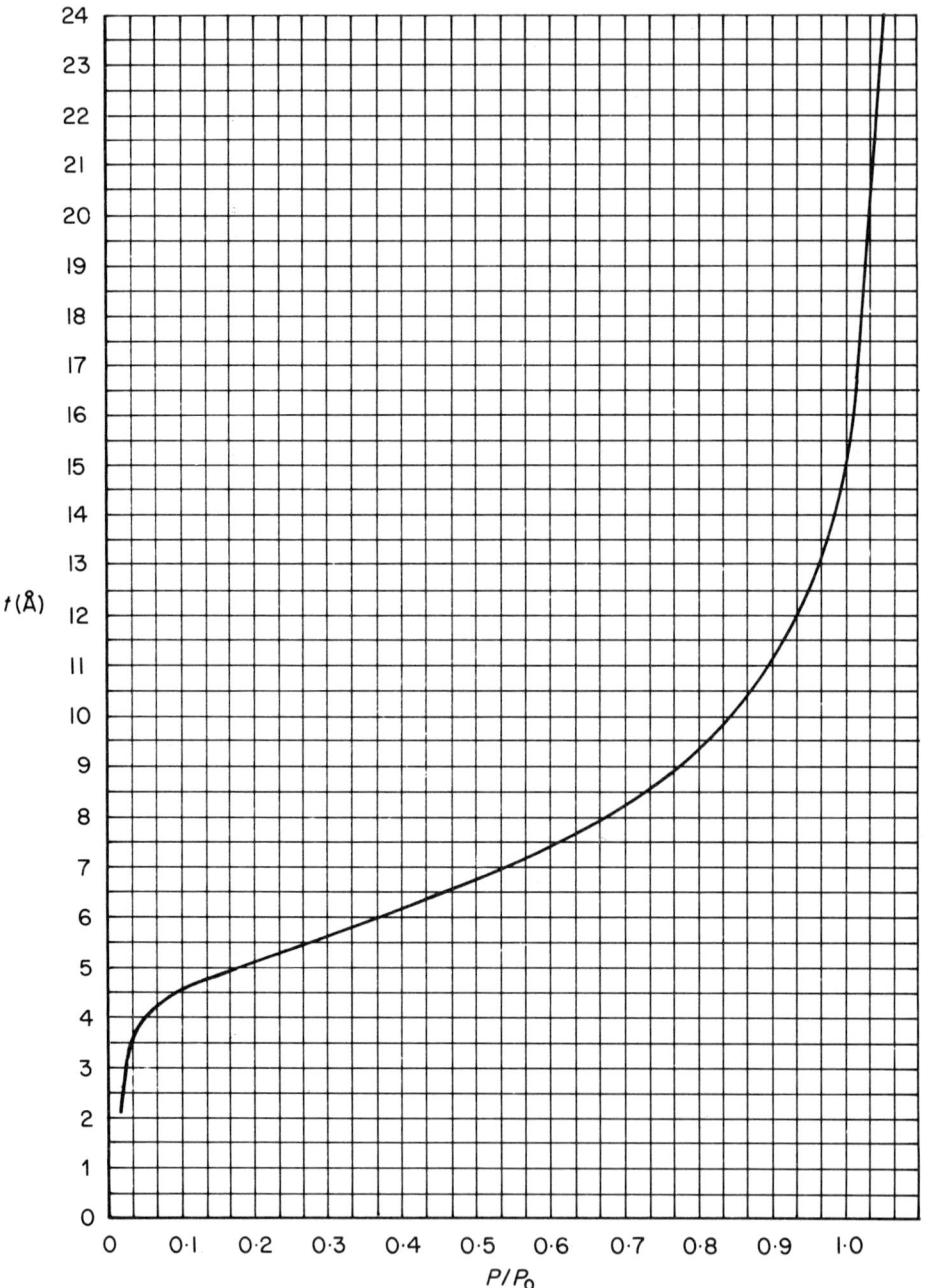

Figure 8.6 Statistical thickness versus relative pressure from the Halsey equation.

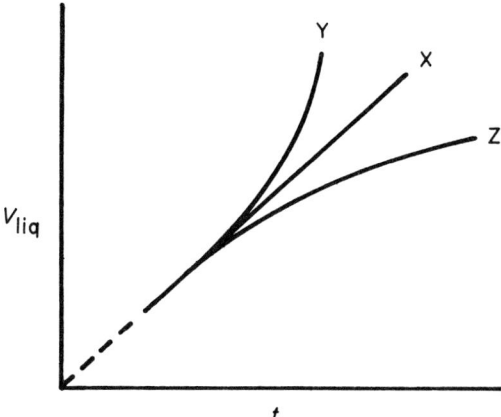

Figure 8.7 $V-t$ curves — basic shapes.

shows that condensation into pores commences where the slope starts to increase. The upward deviation from a straight line reflects the increased uptake due to condensation in larger pores. Curve Z results when narrow pores are filled at low relative pressures by multilayer adsorption, thereby reducing the available surface for continued adsorption.

9
Microporosity

9.1 INTRODUCTION

It is convenient to categorize pores into several size ranges. The largest pore diameters amenable to analysis by the Kelvin equation are about 1000 Å which corresponds to relative pressures near 0.99. Pores with diameters greater than this are termed 'macropores'. Dubinin [74] calls pores in the Kelvin range, from about 15 to 1000 Å, 'transitional' and pores with diameters less than about 15 Å 'micropores'. The failure of V–t plots to pass through the origin has led to the postulation of 'submicropores' [75] with diameters less than about 15 Å. According to the IUPAC convention, micropores are characterized by diameters less than 20 Å, mesopores from 20 to 500 Å, and macropores larger than 500 Å.

Microporous materials exhibit Type I isotherms which are described by equation (4.12), the Langmuir equation. The Langmuir model was developed on the assumption that adsorption was limited to at most one monolayer. However, any factor which can limit the quantity adsorbed to a few monolayers will also produce a Type I isotherm. In the case of very narrow pores, the close proximity of the walls will prevent multilayer adsorption and limit the amount adsorbed.

In view of the intense potential fields in very narrow pores, it would be difficult to determine if the mechanism by which the pores fill is that of adsorption or condensation. However, equation (4.39), a form of the BET equation that treats restricted adsorption, allows for just this kind of limiting adsorption.

9.2 LANGMUIR PLOTS FOR MICROPOROUS SURFACE AREA

A convenient form of the Langmuir equation is

$$\frac{P}{W} = \frac{1}{KW_m} + \frac{P}{W_m} \qquad \text{(cf.4.12)}$$

where P is the adsorbate equilibrium pressure, and W and W_m are the weight adsorbed and monolayer weights, respectively. The term K is a constant discussed in Section 4.1.

For Type I isotherms, a plot of P/W versus P should give a straight line with $1/W_m$ as the slope. The sample surface area S_t is calculated from equation (4.13):

$$S_t = \frac{W_m \bar{N} \mathscr{A}}{\bar{M}} \qquad \text{(cf.4.13)}$$

where, as before, $\mathscr{A}$ is the cross-sectional adsorbate area, $\bar{M}$ is the adsorbate molecular weight and $\bar{N}$ is Avogadro's number.

The fact that a Langmuir plot gives a straight line is not at all indicative of its success. Without an understanding of the processes occurring within the micropores in terms of adsorption or condensation, the Langmuir equation may be a correct mathematical description of the isotherm but the interpretation of the meaning of the monolayer weight W_m remains unresolved.

9.3 EXTENSIONS OF POLANYI'S THEORY FOR MICROPORE VOLUME AND AREA

Polanyi's [76] potential theory of adsorption views the area immediately above an adsorbent's surface as containing equipotential lines which follow the contour of the surface potential. When a molecule is adsorbed, it is considered trapped between the surface and the limiting equipotential plane at which the 'adsorption potential' has fallen to zero. Fig. 9.1 illustrates these equipotential planes. In the diagram, Y represents a pore and X depicts some surface impurity.

According to the potential theory, the volume $\widetilde{V}$, defined by the adsorbent's surface and the equipotential plane E_n, can contain adsorbate in three different conditions depending upon the temperature. Above the critical

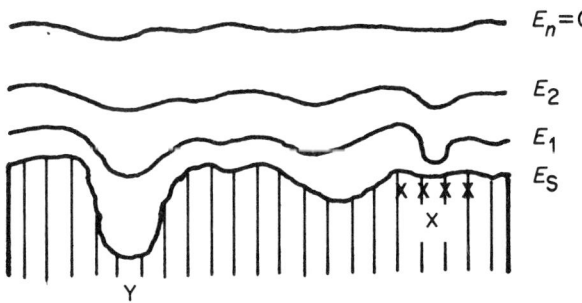

Figure 9.1 Polanyi's potential planes.

temperature, the adsorbate can not be liquified and the gas in the adsorption volume $\tilde{V}$ simply becomes more dense near the surface. At temperatures near, but less than the critical temperature, the adsorbate is viewed as a liquid near the surface and a vapor of decreasing density away from the surface. Substantially below the critical temperature ($T \leq 0.8T_c$), the adsorption volume is considered to contain only liquid. Under the latter conditions one can write

$$\tilde{V} = \frac{W}{\rho} \tag{9.1}$$

where W and ρ are the adsorbate weight and density, respectively.

The potential theory asserts that the adsorption potential, when the adsorbate is in the liquid state, is given by

$$E = RT \ln \frac{P_0}{P} \tag{9.2}$$

According to the preceding equation, E is the isothermal work required to compress the vapor from its equilibrium pressure P to the saturated pressure P_0 of the liquid in the adsorption volume.

Using equations (9.1) and (9.2) both $\tilde{V}$ and E can be calculated from an experimental isotherm. Therefore,

$$\tilde{V} = F(E) \tag{9.3}$$

Plots of $\tilde{V}$ versus E take the form shown in Fig. 9.2 and are called characteristic curves.

If two adsorbates fill the same adsorption volume as shown by the vertical dotted line in Fig. 9.2, their adsorption potentials E and E_0 will differ only because of differences in their molecular properties. Consequently, the ratio of adsorption potentials is assumed by Dubinin [77] to be constant and he calls E/E_0 the 'affinity coefficient' which, for an adsorbate pair, is a measure of their relative affinities for a surface or their adsorbability. Using the adsorption for one vapor as a reference value, say E_0, the ratio of potentials can be written as

$$\frac{E}{E_0} = \beta \tag{9.4}$$

Substitution into equation (9.3) then gives, for the reference vapor

$$\tilde{V} = F\left(\frac{E}{\beta}\right) \tag{9.5}$$

Extensions of Polanyi's theory for micropore volume and area   75

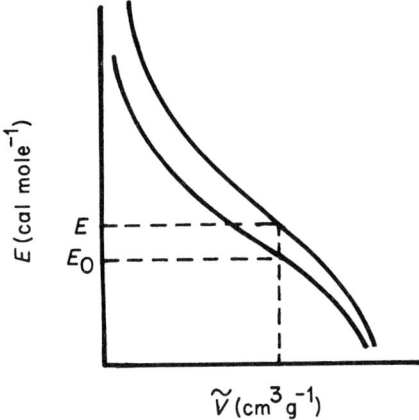

Figure 9.2   Characteristic curves for two vapors.

Using benzene as the reference or standard vapor for which β is taken as unity, Dubinin and Timofeev [78] were able to calculate values of β for other adsorbents.

The similarity in the shapes of the characteristic curves, shown in Fig. 9.2, and the positive side of a Gaussian curve, led Dubinin and Radushkevich [79] to postulate that the fraction of the adsorption volume $\tilde{V}$ occupied by liquid adsorbate at various values of adsorption potentials E can be expressed as a Gaussian function. Thus,

$$\tilde{V} = \tilde{V}_0 e^{-KE_0^2} \tag{9.6}$$

Substituting the value for E_0 from equation (9.4) gives

$$\tilde{V} = \tilde{V}_0 e^{-K(E/\beta)^2} \tag{9.7}$$

where K is a constant determined by the shape of the pore size distribution curve and $\tilde{V}$ is the total adsorption volume or the total microporous volume.

Equation (9.7) is applicable to micropores rather than larger pores because the overlapping potential from the walls of pores only slightly larger than an adsorbate molecule will considerably enhance the adsorption potential

Substituting for E in equation (9.7) using equation (9.2) yields

$$\tilde{V} = \tilde{V} \exp - \left[\frac{K}{\beta^2} RT \left(\ln \frac{P_0}{P} \right)^2 \right] \tag{9.8}$$

which can be rewritten as

$$\log W = \log(\tilde{V}_0 \rho) - 2.303 K \left(\frac{RT}{\beta}\right)^2 \left[\log\left(\frac{P_0}{P}\right)\right]^2 \tag{9.9}$$

where W and ρ are the weight adsorbed and the liquid adsorbate density. Simplification of equation (9.9) gives

$$\log W = \log(\tilde{V}_0 \rho) - K \left[\log\left(\frac{P_0}{P}\right)\right]^2 \tag{9.10}$$

where

$$k = 2.303 K \left(\frac{RT}{\beta}\right)^2 \tag{9.11}$$

A plot of $\log W$ versus $\log[(P_0/P)]^2$ should give a straight line with an intercept of $\log(\tilde{V}_0 \rho)$ from which $\tilde{V}_0$, the micropore volume, can be calculated. Nikolayev and Dubinin [80] found linear plots using relative pressures ranging from 10^{-5} to 10^{-1} on a variety of microporous samples.

Kaganer [81] modified Dubinin's method in order to calculate the surface area within micropores. He assumed that the adsorption potential of the sites is distributed according to a Gaussian function such that

$$\theta = e^{-\bar{k}E^2} \tag{9.12}$$

where θ is the fraction of the surface covered by adsorbate. Combining equations (9.12) and (9.2) in logarithmic form yields

$$\log \theta = -2.303 \bar{k} (RT)^2 \left[\log\left(\frac{P_0}{P}\right)\right]^2 \tag{9.13}$$

The fraction of surface θ, covered by adsorbate, for less than monolayer coverage, is the adsorbed weight W relative to the monolayer weight W_m or

$$\theta = \frac{W}{W_m} \tag{9.14}$$

Equation (9.13) then becomes

$$\log W = \log W_m - \bar{K} \left[\log\left(\frac{P_0}{P}\right)\right]^2 \tag{9.15}$$

where

$$\bar{K} = 2.303 \bar{k} (RT)^2 \tag{9.16}$$

Equation (9.15) is similar to equation (9.10), Dubinin's equation. A plot of $\log W$ versus $(\log P_0/P)^2$ will yield a straight line with an intercept of $\log W_m$ from which the surface area can be calculated by equation (4.13). The linear range of these plots are usually at very low relative pressures, less than 10^{-2}. Kaganer showed excellent agreement between surface areas measured using equation (9.15) and the BET method on a variety of adsorbate–adsorbent systems.

9.4 THE t-METHOD

The utilization of the technique of comparison of an isotherm of a microporous material with a standard Type II isotherm was proposed by de Boer et al. [82] for the determination of micropore volume and surface area, based on the t-curve, a plot of t, the statistical thickness, versus the relative pressure, P/P_0. The t-method employs a composite t-curve obtained from data on a number of nonporous adsorbents with BET C constants similar to those of the microporous samples being tested. The standard t-curve is expressed by the empirical de Boer equation:

$$t(\text{Å}) = \left(\frac{13.99}{\log P_0/P + 0.034}\right)^{\frac{1}{2}} \tag{9.17}$$

The calculation of t from equation (9.17) involves the assumption of hexagonal close packing, that is, that the thickness of a single molecular layer of nitrogen is 3.54 Å. The calculated statistical thickness at various relative pressures is then used to replot the analysis isotherm as a t-curve, a plot of the volume of gas adsorbed versus t. Typical t-plots for both microporous and non-microporous samples are shown in Fig. 9.3.

If the isotherm (A) is identical in shape to the standard isotherm of a nonporous sample (Type II) the t-plot (B) will be a straight line passing through the origin, the slope of which is a measure of surface area, according to equation (8.36):

$$S_t = \frac{V_{\text{liq}}}{t} \times 10^4 \tag{cf.8.36}$$

where S_t is the total surface area and V_{liq} is the adsorbed liquid volume.

When a microporous material is mixed with the nonporous sample, the adsorption isotherm (C) will show an increased uptake of gas at low relative pressures, the t-plot (D) will be linear and when extrapolated to the adsorption axis will show a positive intercept, equivalent to the micropore volume. The slope of the straight line in D is proportional to the external surface area since all micropores have already been filled by capillary condensation. The

78 Microporosity

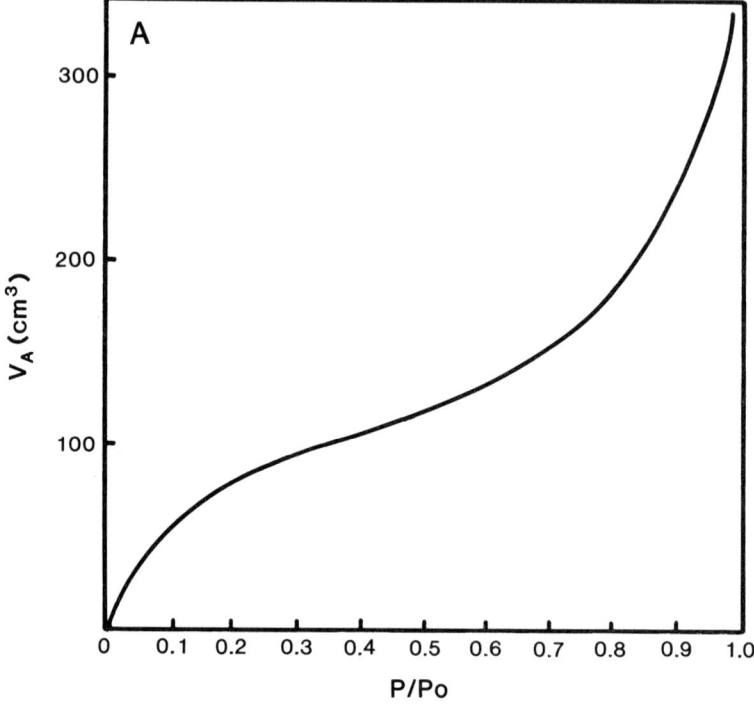

Figure 9.3A Representative isotherms and V_A-t plots: (A) Standard Type II isotherm; (B) t-plot from Type II isotherm; (C) Type II isotherm + microporous sample; (D) t-plot from isotherm C; (E) Isotherm of a microporous material; (F) t-plot from E; (G) Type IV isotherm; (H) t-plot from G; (I) Type IV isotherm + microporous material; (J) t-plots.

adsorption isotherm E is typical of a sample having only micropores. The corresponding t-plot (F) is interpreted in the same manner as in D.

A typical Type IV isotherm G, indicative of the presence of mesopores, would result in the t-plot shown in H, provided no micropores are present.

If microporous material is added to the mesoporous sample the resulting adsorption isotherm I would show capillary condensation into micropores at low relative pressures. The corresponding t-plots shown in Fig. 9.3 J are illustrative of micropores in the presence of mesopores. The initial slope of the V_A-t curve corresponds to small values of t which represents an adsorbed film within large pores and complete filling of smaller pores. Those micropores smaller in diameter than the adsorbate molecule cannot contribute to this gas uptake. Therefore, from the initial slope of the V_A-t curve the total surface area of the sample can be obtained using equation (8.36). This area should agree with that calculated from the BET method for porous materials.

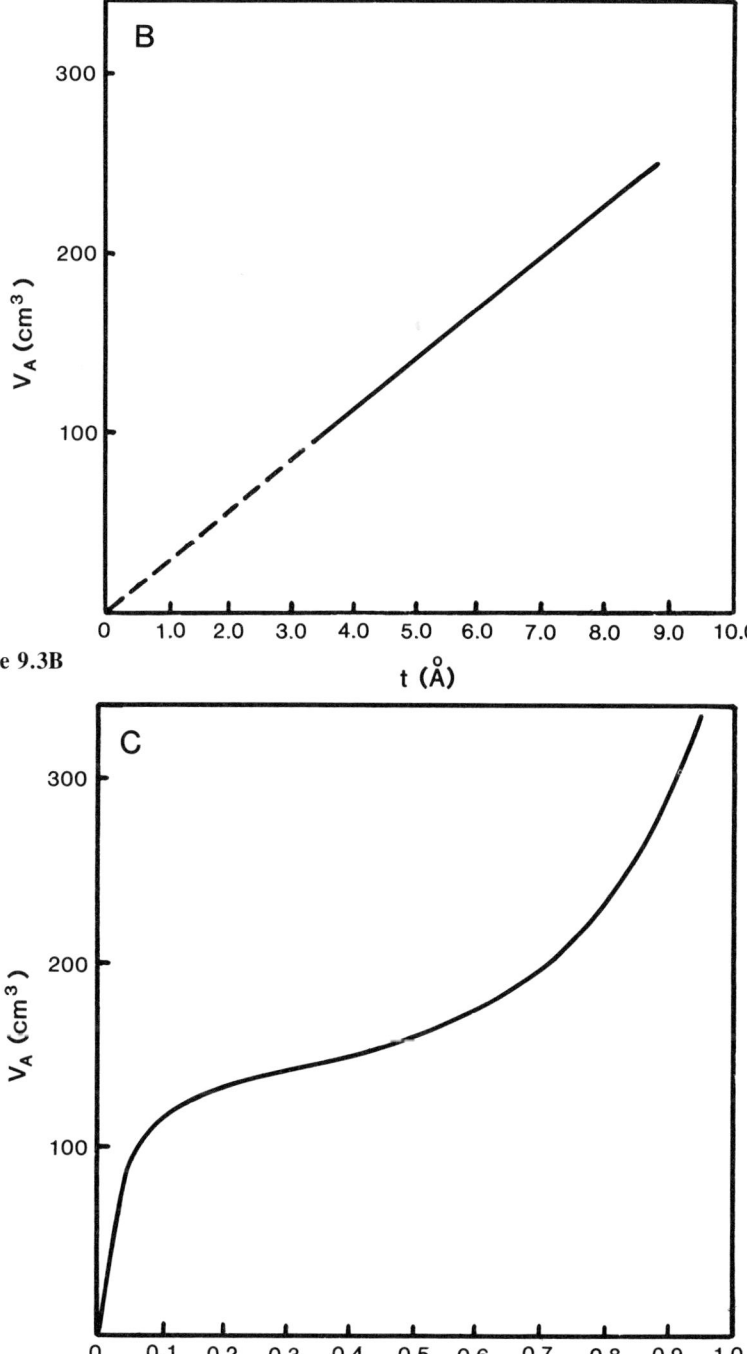

Figure 9.3B

Figure 9.3C

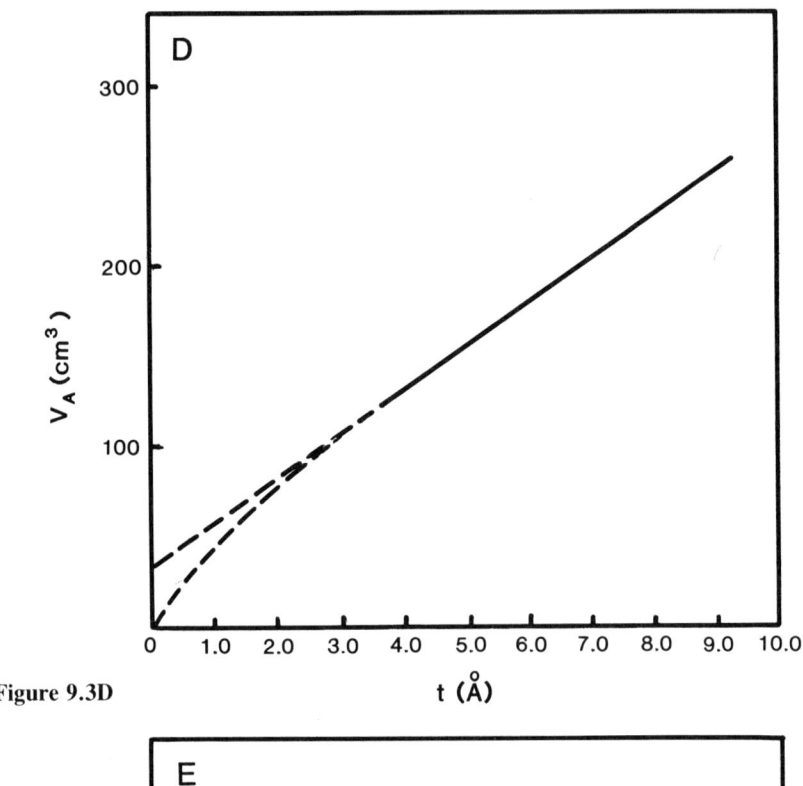

Figure 9.3D

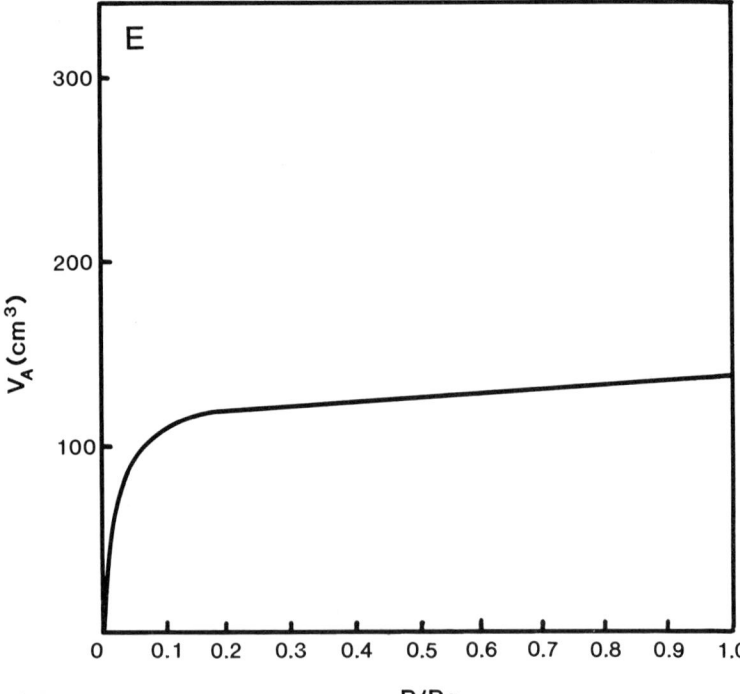

Figure 9.3E

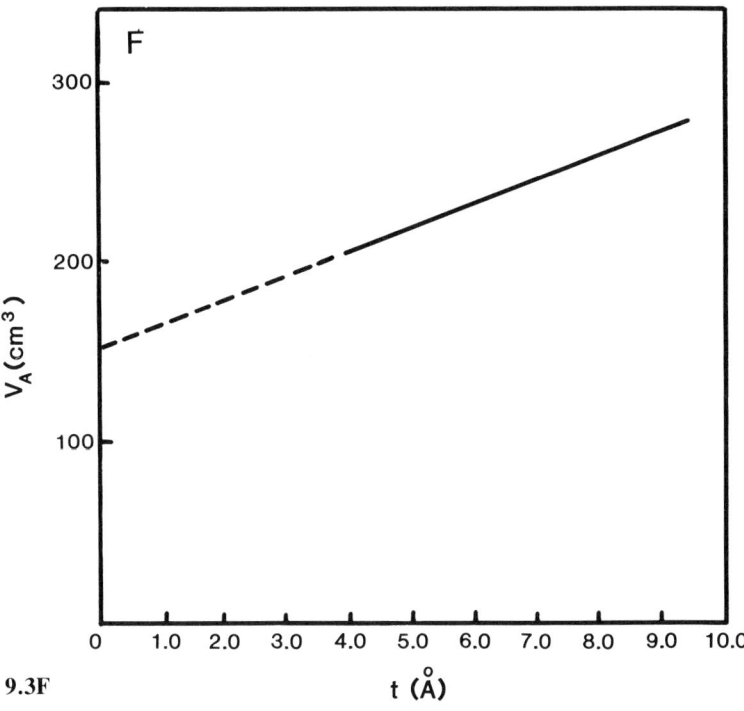

Figure 9.3F

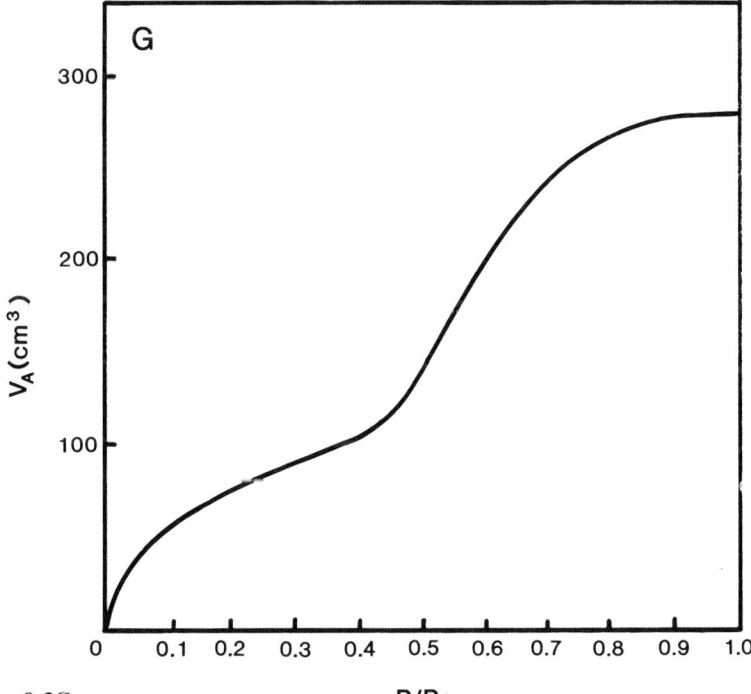

Figure 9.3G

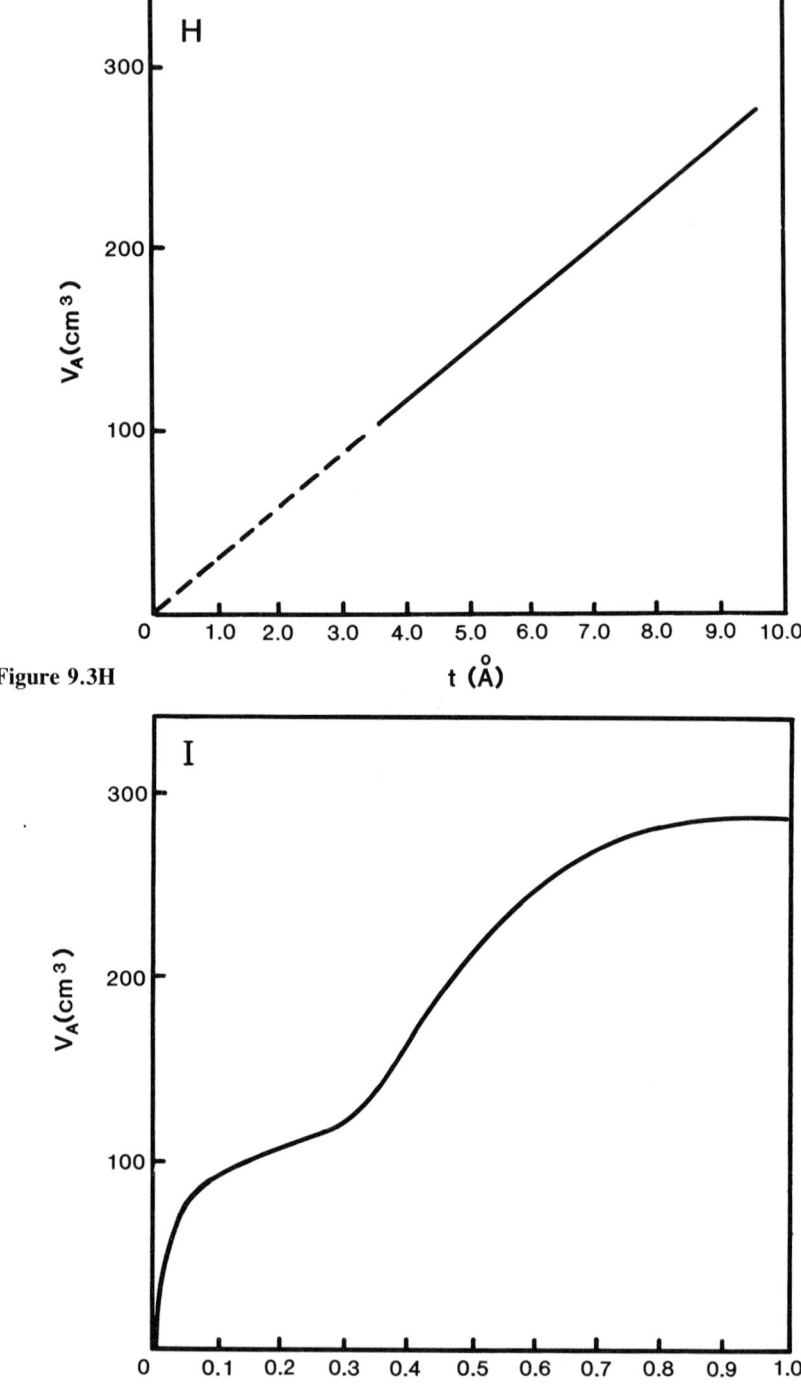

Figure 9.3H

Figure 9.3I

The t-method 83

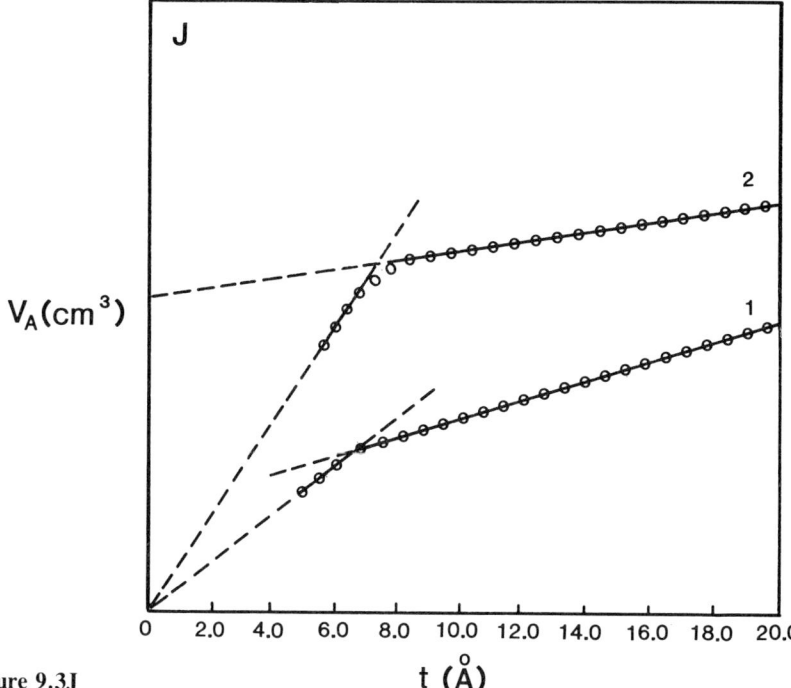

Figure 9.3J

The surface area of the wide pores is similarly obtained from the slope of the upper linear portion of the *t*-plot. This area represents the build-up of a statistical thickness in all pores except the micropores which are presumed filled at higher *t* values. There the difference between these two surface areas is the surface area of the micropores only. The region between the two linear portions represents the transition that occurs as the micropores become filled while multilayer adsorption continues to occur in the larger pores. Thus, after converting the gas volume to the corresponding liquid volume the micropore area can be calculated by applying equation (8.36), as follows:

$$S_{micro} = \left(\frac{V_{liq}}{t}\right)_{lower} - \left(\frac{V_{liq}}{t}\right)_{upper} \times 10^4 \qquad (cf.8.36)$$

In the absence of sufficient data at low relative pressures, the total surface area from the lower linear portion of the *t*-curve cannot be calculated easily. Because the total area should be the same as the BET surface area, the micropore surface area can be calculated from:

$$S_{micro} = S_{BET} - (V_{liq}/t)_{upper} \times 10^4 \qquad (9.18)$$

The abrupt break in the two linear parts of the *t*-plot shown in Fig. 9.3 J-1 indicates the presence of a group of micropores in a narrow pore size range,

84 Microporosity

whereas the curvature between the two linear portions of J-2 is an indication of a wider distribution of micropores.

The *t*-method has been utilized by Johnson [83] to estimate the zeolite content of catalysts by nitrogen adsorption.

The micropore volume of a sample is also obtained from a *t*-plot by extrapolating the upper linear part of the curve to the volume axis. The intercept of this line, after conversion to liquid volume, gives the micropore volume.

9.5 THE α_s-METHOD

Another method for estimating micropore volume without the assumption of a knowledge of the adsorbate statistical thickness is the α_s-method developed by Gregg and Sing [84]. The standard isotherm in this method is a plot of the amount of gas adsorbed normalized by the amount of gas adsorbed at a fixed relative pressure, usually at $P/P_0 = 0.4$, versus P/P_0 (see Fig. 9.4). The normalized term $V_{ads}/V_{ads}^{0.4}$ is α_s.

The α_s plot is subsequently obtained by plotting the volume of gas adsorbed by a test sample versus α_s in the same way as one produces a *t*-plot. The

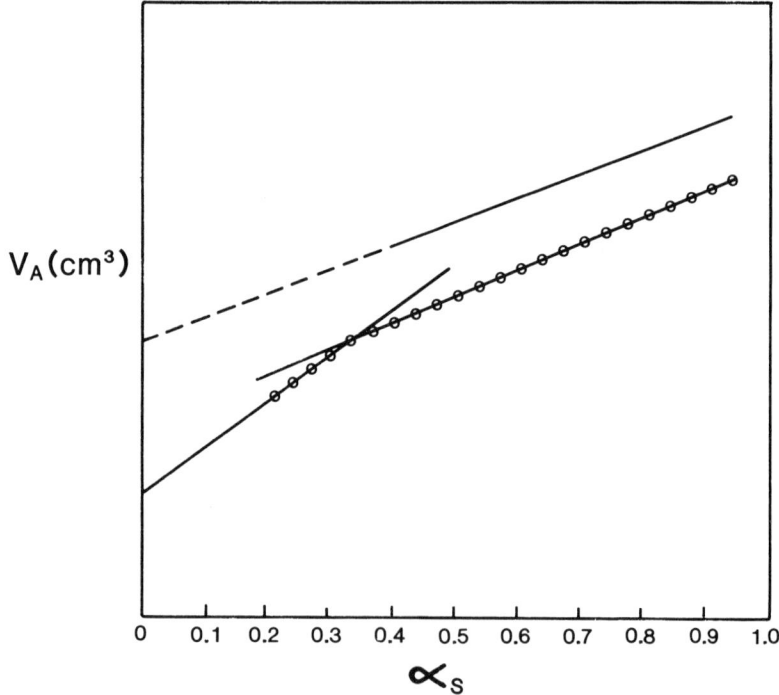

Figure 9.4 α_s-plot.

estimation of micropore volume from an α_s plot, as in the *t*-method, involves extrapolation of the plots to the Y axis.

Since the α_s-method does not assume any value for the thickness of an adsorbed layer, the calculation of surface area is accomplished by relating the slope of the α_s-plot of the test sample to the slope of the corresponding plot for the standard sample of known surface area. One advantage of this fact is that the α_s-method can be used with any adsorbate gas, while the *t*-method is restricted to nitrogen.

9.6 THE MICROPORE ANALYSIS METHOD

Mikhail, Brunauer and Bodor [85] proposed an extension of de Boer's *t*-method for the analysis of micropores which offers several advantages. These include the ability to obtain the micropore volume, surface area, and their distributions from one experimental isotherm. Data for the MP (micropore analysis) method need not be measured at the very low pressures needed for the Dubinin and Kaganer theories. The method assumes no site energy or adsorption volume distribution functions as required by the Kaganer and Dubinin theories. Also, the MP method is applicable to adsorbents containing macro-, transitional (mesopores) and micropores with self-termination when the micropore analysis is completed.

When measuring mesopores, the statistical thickness *t* is used as a correction to allow for desorption from the adsorbed film. However, the value of *t* is much more critical when measuring micropores because in the MP method, *t* is the actual measure of the pore size.

The usual method of measuring the statistical thickness is to divide the liquid volume of nitrogen adsorbed, by the BET surface area

$$t = \frac{V_{liq}}{S_{BET}} \times 10^4 (\text{Å}) \qquad \text{(cf.8.36)}$$

The assumption usually made is that the ratio V_{liq}/S_{BET} has the same value at a given relative pressure independent of the solid. A plot therefore of *t* versus P/P_0 should give the same curve for any non-porous solid (see Fig. 8.6). In fact, plots of the number of adsorbed layers versus P/P_0 show some discrepancies which for the analysis of large pores is not significant. Therefore, the Halscy equation can be used for the statistical thickness in that application. However, for micropore analysis, a statistical thickness must be taken from a *t* versus P/P_0 curve that has approximately the same BET *C* value as the test sample. The unavailability of *t* versus P/P_0 plots on numerous surfaces with various *C* values would make the MP method of passing interest were it not for the fact that *t* can be calculated from equation (8.36). This implies that surface area can be accurately measured on microporous samples.

86 Microporosity

Brunauer [86] points out that in most instances the BET equation does correctly measure the micropore surface area.

As stated earlier, the C value in micropores will be large due to the overlapping wall potentials. Under these circumstances, the surface will be covered well over 90 per cent by stacks of adsorbate not in excess of two molecules in depth as shown by equation (4.45) and Table 4.1. Therefore, the close proximity of the walls offer no special condition which is not already allowed for by the BET theory.

To illustrate the MP method, consider the isotherm shown in Fig. 9.5. The volumes of adsorbed gas are converted to liquid volumes from which t is calculated using equation (8.36). The V–t plot shown in Fig. 9.6 is then constructed using relative pressure intervals of 0.05.

In the example shown, the t values were not calculated from equation (8.36) but were taken from a t versus P/P_0 plot prepared by de Boer et al. [87] on a sample with a similar BET C value. By choosing t values from a material with a similar but not identical C value, the surface area nevertheless agreed to within 1.4% of the BET measured area.

According to Fig. 8.7, it is evident that the shape of Fig. 9.5 is associated with microporosity. The slope of the linear portion of the curve, the origin through the first four points, is 0.0792 which gives a micropore surface area of

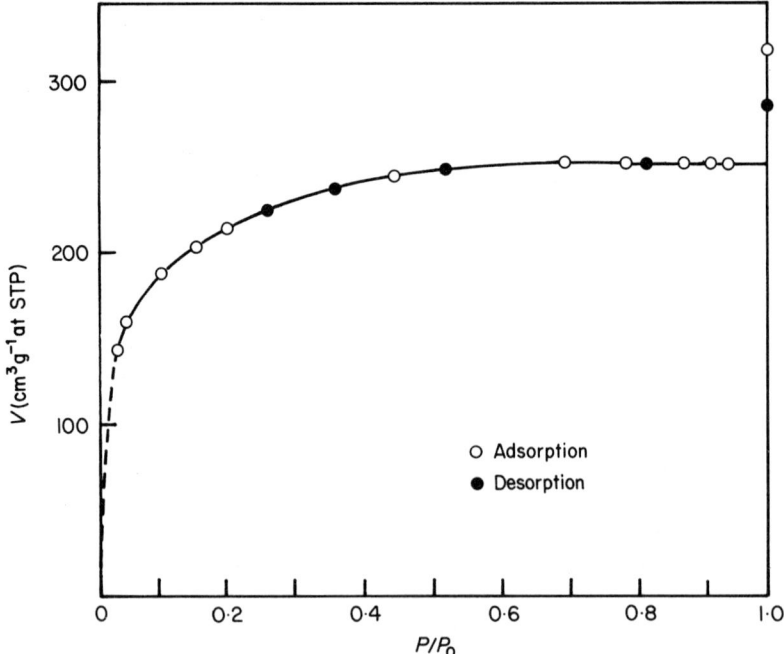

Figure 9.5 Isotherm of N_2 on silica gel, Davidson 03 at 77.3 °K (from Ref. 85).

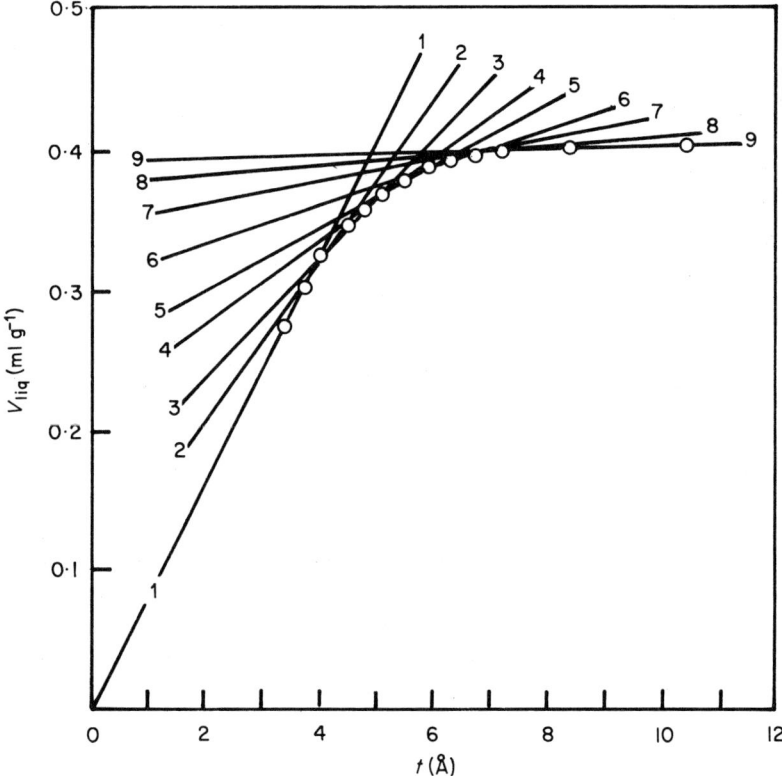

Figure 9.6 $V-t$ curve from Fig. 9.5 (from Ref. 85).

792 m²/g when calculated from equation (8.36). Straight line 2, drawn tangent to the curve between $t = 4$ and 4.5 Å, exhibits a slope of 0.0520. The area of all the pores remaining unfilled by adsorbate is 520 m²/g and the surface area of pores in the range of thickness from 4 to 4.5 Å is $792 - 520 - 272$ m²/g. The third line gives a slope of 0.0360 between t values of 4.5 and 5 Å and the area of pores in this range is $520 - 360 = 160$ m²/g. The calculation is continued in this manner until there is no further decrease in the slope of the $V-t$ plot which means that all the micropores are filled.

The calculation of pore volume is carried out in an equally straightforward manner. For example, the volume of pores is given by

$$V = 10^{-4}(S_1 - S_2)\frac{(t_1 + t_2)}{2} \text{ cm}^3/\text{g} \tag{9.19}$$

Thus, for the first group of pores the volume is

88 Microporosity

$$V - 10^{-4}(792 - 520)\frac{(4.0 + 4.5)}{2} = 0.1156\,\text{cm}^3/\text{g} \qquad (9.20)$$

The micropore data for the isotherm shown in Fig. 9.5 is illustrated in Table 9.1, in which S_i is the total surface area.

The exact pore shape is usually unknown and cylindrical pores are generally assumed. Mikhail, Brunauer and Bodor [85] show in their paper that equation (9.19) is equally valid for parallel plate or cylindrical pores and that the mean hydraulic radius r_h, in Table 9.1, is the same as the separation between plates or the cylindrical radius.

9.7 TOTAL MICROPORE VOLUME AND SURFACE AREA

Both de Boer's *t*-method and Brunauer's MP method are based on the assumption that the BET measured surface area is valid for micropores. Shields and Lowell [88], using this same assumption, have proposed a method for the determination of the micropore surface area using mercury porosimetric data. The surface area of micropores is determined as the difference between the BET surface area and that obtained from mercury porosimetry (see Section 12.4). Since mercury porosimetry is capable of measuring pore sizes only as small as approximately 18 Å radii, this technique affords a means

Table 9.1

Pore group	S_{i+1} (m²/g)	$S_i - S_{i+1}$[†] (m²/g)	mean r_h (Å)	V_i (cm³/g)
1	520	272	4.25	0.1156
2	360	160	4.75	0.0760
3	280	80	5.25	0.0420
4	200	80	5.75	0.0460
5	140	60	6.25	0.0375
6	80	60	6.75	0.0405
7	20	60	7.25	0.0435
8	10	10	7.75	0.0077
		$\Sigma S_i = 782$		

BET area = 793 m²/g
V–t area 782 m²/g
Total Pore volume = 0.4034
MP pore volume = 0.4088

[†] The authors attributed the difference between the V–t and BET areas to surfaces that did not lie within pores.

of calculating the surface area of all pores of radii between 3.5 Å (the diameter of a nitrogen molecule) and 18 Å. Similarly, Shields and Lowell [88] suggested a method for the determination of the total volume of micropores by a combination of techniques. The difference between the sample volumes measured by mercury porosimetry and helium pycnometry (see Chapter 22) is the volume occupied by the micropores.

10
Theory of wetting and capillarity for mercury porosimetry

10.1 INTRODUCTION

The experimental method of mercury porosimetry for the determination of the porous properties of solids is dependent on several variables. One of these is the wetting or contact angle between mercury and the surface of the solid.

When a liquid is placed in contact with a surface of a porous solid the question arises as to whether it will penetrate into the pores. The answer must be pursued in the realm of capillarity which deals with the equilibrium geometries of liquid–solid interfaces and the angle of contact between the liquid and the pore wall.

In the absence of gravity or other external forces, a liquid will assume a spherical shape that possesses the minimum area to volume ratio of all geometric forms. If the sphere is distorted by any external force, molecules must be brought from the interior to the surface in order to provide for the increased surface area. This process will require that work be expended in order to raise the potential energy of a molecule when the number of stabilizing interactions with neighboring molecules is reduced upon reaching the surface. The work expended will raise the free energy G of the liquid. That part of the total free energy change ΔG which is altered is the free surface energy G^s. Therefore, the change of free surface energy ΔG^s is the net work required to alter the surface area of a substance. Since spontaneous processes are associated with a decrease in free energy, in the absence of external forces, liquids will spontaneously assume a spherical shape in order to minimize their exposed area and thereby their free surface energy. The spontaneous coalescence of two similar liquid droplets into one larger drop when brought into contact is a dramatic demonstration of the free surface energy decrease brought about by the decrease in surface area by the formation of a single larger drop.

The surface tension, γ, of a substance is identical to the free surface energy G^s per unit area and is the work required to alter the surface area by one square centimeter. Therefore, γ has the dimensions of energy per unit area.

According to the preceding argument, a bubble should collapse in order to minimize its free surface energy. In the process of shrinking, however, the gas pressure within the bubble will increase preventing any further radius reduction. When the bubble radius decreases from r to $r - dr$, the free energy decreases by

$$-dG = 8\pi r \gamma dr \tag{10.1}$$

When the bubble shrinks, the volume change is $4\pi r^2 dr$. The gas within the bubble undergoes compression while the external atmosphere undergoes expansion. The net work associated with the compression and expansion processes is given by

$$-dW = (P_{int} - P_{ext}) 4\pi r^2 dr \tag{10.2}$$

The internal pressure, P_{int}, is greater than the external pressure, P_{ext}, since it supports both the external pressure and the tendency for the surface area of the film to decrease. Since the work performed is equal to the free energy difference, one obtains from equations (10.1) and (10.2) at mechanical equilibrium

$$(P_{int} - P_{ext}) = \frac{2\gamma}{r} \tag{10.3}$$

or

$$\Delta P = \frac{2\gamma}{r} \tag{10.4}$$

Equation (10.4) dictates that the smaller the bubble radius the greater will be the pressure difference across the wall. Thus, a large and small bubble each exposed to the same external pressure will result in a greater internal pressure within the smaller bubble. A vivid demonstration of this occurs when a balloon is inflated. The lung pressure required decreases rapidly as the balloon radius increases until the elastic limit is approached.

10.2 YOUNG AND LAPLACE EQUATION

Equation (10.4) is a special case of a more general concept represented by the Young [89] and Laplace [90] equation. A sphere possesses a constant radius of curvature. For an area element belonging to a nonspherical curved surface, there can exist two radii of curvature (r_1 and r_2). If the two radii of curvature are maintained constant while an element of the surface is stretched along the x-axis from x to $x + dx$ and along the y-axis from y to $y + dy$ the work W_1 performed will be

$$W_1 = \gamma[(x + dx)(y + dy) - xy] \tag{10.5}$$

which, ignoring the product of differentials, reduces to

$$W_1 = \gamma(xdy + ydx) \tag{10.6}$$

If the area element is stretched due to an increase in internal pressure relative to the external pressure there will also be displacement along the z-axis as the surface expands. The work performed will be

$$W_2 = \Delta P(x + dx)(y + dy) dz \tag{10.7}$$

Again, neglecting the product of differentials, equation (10.7) reduces to

$$W_2 = \Delta P xy dz \tag{10.8}$$

Because the two radii of curvature r_1 and r_2 remain unchanged, one can write

$$\frac{x + dx}{r_1 + dz} = \frac{x}{r_1} \tag{10.9}$$

and

$$\frac{y + dy}{r_2 + dz} = \frac{y}{r_2} \tag{10.10}$$

Equations (10.9) and (10.10) reduce to

$$dx = \frac{xdz}{r_1} \tag{10.11}$$

and

$$dy = \frac{ydz}{r_2} \tag{10.12}$$

Substitution of the values in equations (10.11) and (10.12) for dx and dy into equation (10.6) yields

$$W_1 = \gamma\left(\frac{1}{r_1} + \frac{1}{r_2}\right) xy dz \tag{10.13}$$

At mechanical equilibrium W_1 must equal W_2, leading to

$$\Delta P = \gamma \left(\frac{1}{r_1} + \frac{1}{r_2} \right) \tag{10.14}$$

The Young and Laplace equation, equation (10.14), reduces to equation (10.4) for the special case of a sphere with r_1 equal to r_2. For a bubble, the right-hand side of equations (10.14) and (10.4) should be multiplied by 2 to allow for the fact that there are two surfaces being stretched, the interior and the exterior.

10.3 WETTING OR CONTACT ANGLES

The affinity of a liquid for a solid surface is usually described as wetting. When a liquid spreads spontaneously along a solid surface it is said to wet the surface. If the liquid in the form of a drop remains stationary and appears spherical it is nonwetting. Fig. 10.1 illustrates wetting and nonwetting liquids on a solid surface. A measure of the degree of wetting is given by the wetting or contact angle, θ, the measurement of which is discussed in Chapter 21. The wetting angle is greater or less than 90° for nonwetting and wetting liquids, respectively.

A drop of liquid at rest on a solid surface is under the influence of three forces or tensions. As shown in Fig. 10.2, the circumference of the area of

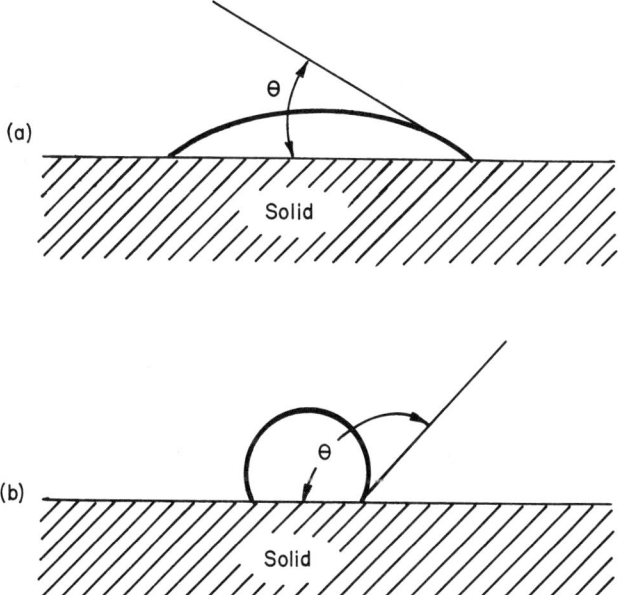

Figure 10.1 Wetting and nonwetting liquids on a solid surface: (a) Wetting, $\theta < 90°$; (b) Nonwetting, $\theta > 90°$.

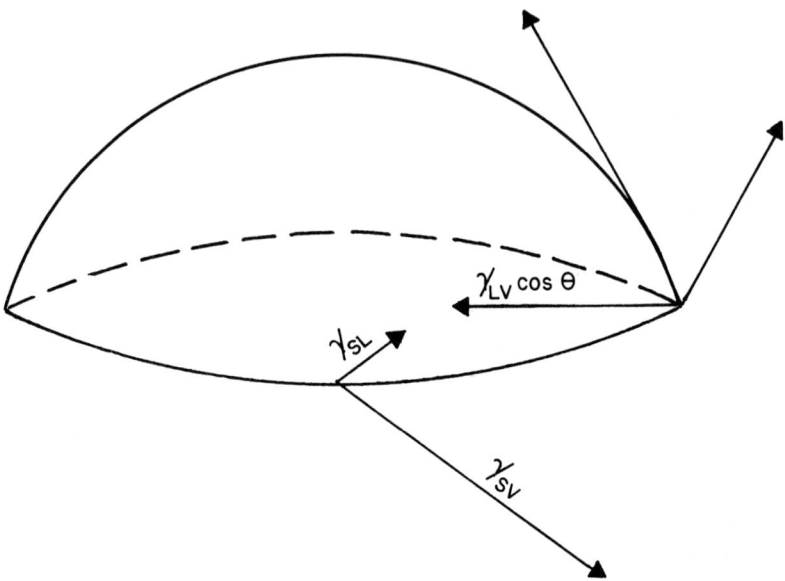

Figure 10.2 Interfacial tensions.

contact of a circular drop is drawn toward the center of the drop by the solid–liquid interfacial tension, γ_{SL}. The equilibrium vapor pressure of the liquid produces an adsorbed layer on the solid surface that causes the circumference to move away from the drop center and is equivalent to a solid–vapor interfacial tension, γ_{SV}. The interfacial tension between the liquid and vapor, γ_{LV}, essentially equivalent to the surface tension γ of the liquid, acts tangentially to the contact angle θ, drawing the liquid toward the drop center with its component $\gamma_{LV}\cos\theta$.

At mechanical equilibrium the tensions or forces cancel to yield

$$\gamma_{SV} = \gamma_{SL} + \gamma_{LV}\cos\theta \tag{10.15}$$

Equation (10.15), derived by Young [89] and Dupre [91], establishes the criteria for wettability. The contact angle is determined from the values of the three interfacial tensions. When γ_{SV} exceeds γ_{SL} then $\cos\theta$ must be positive with $\theta < 90°$. When γ_{SL} is greater than γ_{SV} the $\cos\theta$ term is negative and $\theta > 90°$. Complete wetting of a surface occurs at a contact angle of 0° and total nonwetting at 180°. Wetting and nonwetting are general terms usually associated with contact angles less than or greater than 90°, respectively.

10.4 CAPILLARITY

When a capillary is immersed under the surface of a liquid, the liquid will either rise in the capillary (Fig. 10.3a) or be depressed below the surface of the external liquid (Fig. 10.3b).

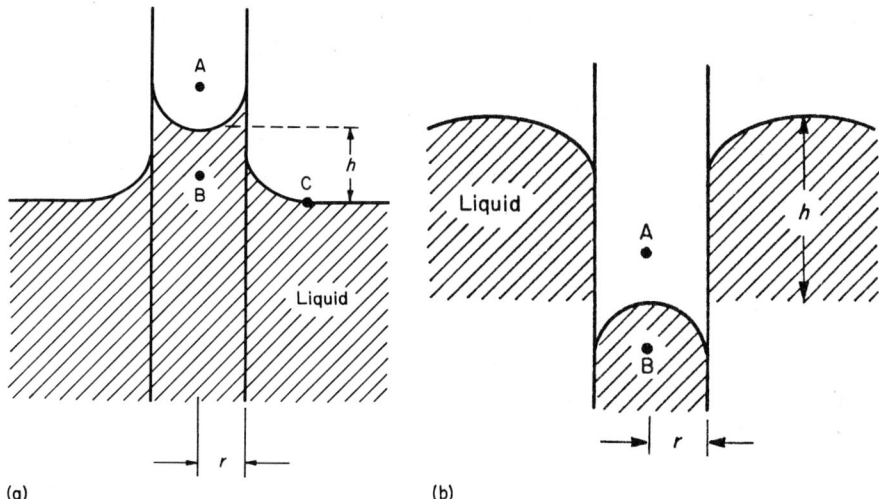

Figure 10.3 (a) Capillary rise, $\theta < 90°$; (b) Capillary depression, $\theta > 90°$.

As illustrated in Fig. 10.3a, the pressure above the meniscus within a capillary of radius r at point A is the same as that at point C at the surface of the liquid level external to the capillary. The small pressure difference due to the gravitational difference in gas densities is neglected. According to equation (10.4) the pressure on the concave side of a curved surface at point A must be in excess of that on the convex side at point B. Because the pressure at point B is less than that at A the liquid will rise within the capillary until the sum of the pressure at point B and the pressure head induced by the liquid rising to height h is equal to the external pressure. At equilibrium then

$$P_A = P_B + hg(\rho_l - \rho_g) \tag{10.16}$$

where P_A and P_B are the pressures at points A and B, ρ_l and ρ_g are the densities of the liquid and the gas that it displaces in the capillary, and g is the gravitational constant. Equation (10.16) is more conveniently expressed as

$$P_A - P_B = \Delta P = hg(\rho_l - \rho_g) \tag{10.17}$$

Combining equations (10.17) and (10.4) yields

$$\frac{2\gamma}{r} = hg(\rho_l - \rho_g) \tag{10.18}$$

which is often used to describe the surface tension of a liquid by the capillary rise method.

When a liquid does not wet the walls of a capillary, as shown in Fig. 10.3b, the concave side of the meniscus is within the liquid at point B, which is at a higher pressure than the gas immediately above the surface. As described by equation (10.18), the liquid is depressed a distance h below the level of the external liquid.

Equation (10.18) was derived for capillary rise or depression assuming complete wetting, that is, $\theta = 0°$. In the case of contact angles greater than $0°$ and less than $180°$, equation (10.18) must be modified. As liquid moves up the capillary during capillary rise the solid–vapor interface disappears and the solid–liquid interface appears. The work required for this process is

$$W = (\gamma_{SL} - \gamma_{SV}) \Delta A \tag{10.19}$$

where ΔA is the area of the capillary wall covered by liquid as it rises. From the Young and Dupre equation, equation (10.15), and equation (10.19), it can be seen that

$$W = -(\gamma_{LV} \cos\theta) \Delta A \tag{10.20}$$

in which γ_{LV} is the surface tension γ of the liquid. The work required to raise a column of liquid a height h in a capillary of radius r is identical to the work that must be performed to force it out of the capillary. When a volume V of liquid is forced out of the capillary with gas at a constant pressure above ambient, ΔP_{gas}, the work is given by

$$W = V \Delta P_{gas} \tag{10.21}$$

Combining equations (10.20) and (10.21) yields

$$\Delta P_{gas} V = (\gamma \cos\theta) \Delta A \tag{10.22}$$

where ΔP_{gas} is the excess pressure above that of the external pressure and V is the volume of liquid displaced. If the capillary is circular in cross-section, the terms V and ΔA are given by $\pi r^2 \ell$ and $2\pi r \ell$, respectively. Substituting these values into equation (10.22) gives

$$\Delta P r = -2\gamma \cos\theta \tag{10.23}$$

10.5 WASHBURN EQUATION

Equation (10.23) was first derived by Washburn [92] and is the operating equation in mercury porosimetry. For wetting angles less than $90°$, $\cos\theta$ is positive and ΔP is negative, indicating that pressure greater than ambient must be applied to the top of a liquid column in a capillary to force the liquid out. When θ is greater than $90°$, resulting in capillary depression, $\cos\theta$ is negative and pressure greater than ambient must be applied to the liquid in the reservoir, in which the capillary is immersed, to force the liquid into the capillary. When $\theta = 0°$, equation (10.23) reduces to equation (10.4) where ΔP is identical to the pressure within a bubble in excess of ambient.

According to the Washburn equation (10.23) a capillary of sufficiently small radius will require more than one atmosphere of pressure differential in order for a nonwetting liquid to enter the capillary. In fact, a capillary with a radius of 18 Å (18×10^{-10} m) would require nearly 414 MPa (60 000 psi) of pressure before mercury would enter — so great is the capillary depression. The method of mercury porosimetry requires evacuation of the sample and subsequent pressurization to force mercury into the pores. Since the pressure difference across the mercury interface is then the applied pressure, equation (10.23) reduces to

$$Pr = -2\gamma \cos\theta \tag{10.24}$$

An alternate derivation of the Washburn equation can be pursued as follows. For a pore of circular cross-section with radius r the surface tension acts to force a non-wetting liquid out of the pore. The force developed due to interfacial tensions is the product of the surface tension γ of the liquid and the circumference ($2\pi r$) of the pore, that is,

$$F = 2\pi r\gamma \tag{10.25}$$

Since the interfacial tension acts tangent to the contact angle θ, the component of force driving mercury out of the pore becomes

$$F_{out} = 2\pi r\gamma \cos\theta \tag{10.26}$$

The force driving mercury into the pore can be expressed as the product of the applied pressure P and the cross-sectional area of the pore directed opposite to F_{out}, that is,

$$-F_{in} = P\pi r^2 \tag{10.27}$$

Equating (10.25) and (10.26) at equilibrium gives

$$Pr = -2\gamma \cos\theta \tag{cf.10.24}$$

Another approach to the Washburn equation involves the work required to force mercury into a pore. Because of its high surface tension, mercury tends not to wet most surfaces and must be forced to enter a pore. When forced under pressure into a pore of radius r and length ℓ an amount of work W is required which is proportional to the increased surface exposed by the mercury at the pore wall. Therefore, assuming cylindrical pore geometry

$$W = 2\pi r \ell \gamma \tag{10.28}$$

Since mercury exhibits a wetting angle greater than 90° and less than 180° on all surfaces with which it does not amalgamate, the work required is reduced by $\cos\theta$ and equation (10.28) becomes

$$W_1 = 2\pi r \ell \gamma \cos\theta \tag{10.29}$$

When a volume of mercury, ΔV, is forced into a pore under external pressure P an amount of work W_2 given as follows, is performed

$$W_2 = -P\Delta V = -P\pi r^2 \ell \tag{10.30}$$

where the negative sign implies a decreasing volume. At equilibrium equations (10.29) and (10.30) are combined to give

$$Pr = -2\gamma \cos\theta \tag{cf.10.24}$$

Because the product Pr is constant, assuming constancy of γ and θ, equation (10.24) dictates that as the pressure increases mercury will intrude into progressively narrower pores.

11
Interpretation of mercury porosimetry data

11.1 APPLICATIONS OF THE WASHBURN EQUATION

The experimental method employed in mercury porosimetry, discussed more extensively in Chapter 21, involves the evacuation of all gas from the volume containing the sample. Mercury is then transferred into the sample container while under vacuum. Finally, pressure is applied to force mercury into the interparticle voids and intraparticle pores. A means of monitoring both the applied pressure and the intruded volume are integral parts of all mercury porosimeters.

The volume forced into the pores is usually monitored in a penetrometer, which is the calibrated precision stem of a glass cell, containing the sample and filled with mercury. As intrusion occurs, the level in the stem decreases. Ritter and Drake [93] used a resistance wire coaxial in the capillary stem to monitor the volume changes. They also developed one of the earliest high pressure porosimeters and measured the contact angle between mercury and a variety of materials, which they found to be between 135 and 142° with an average value of 140°. Using this value for θ and 0.480 N/m (480 dyne cm^{-1}) for γ, at room temperature, converts equation (10.24) into the following expressions using pressure in megapascals (MPa) or kilograms per square centimeter:

$$r = \frac{0.736}{P} \qquad (11.1)$$

where P is in MPa, and r in micrometers and

$$r = \frac{1530}{P} \qquad (11.2)$$

where P is in kg cm^{-2}, and r in micrometers.

100 *Interpretation of mercury porosimetry data*

11.2 INTRUSION–EXTRUSION CURVES

A plot of the intruded (or extruded) volume of mercury versus pressure is sometimes called a porogram. The authors will use the terms 'intrusion curve' to denote the volume change with increasing pressure and 'extrusion curve' to indicate the volume change with decreasing pressure. Fig. 11.1 shows a typical porosimetry curve of cumulative volume plotted versus both pressure (bottom abscissa) and radius (top abscissa). The same data plotted on semilog paper is illustrated in Fig. 11.2.

As shown in Fig. 11.1, the initial intrusion at very low pressure is due to penetration into the interparticle voids when the sample is a powder. The slight positive slope between points A and B on the intrusion curve results from continued filling of the torroidal volume [94–98] between the contacting particles. As the pressure is increased, mercury will penetrate deeper into the narrowing cavities between the particles. Depending on the size, size distribution, shape and packing geometry of the particles, there will exist some interparticle voids of various dimensions and shapes which will progressively fill as the pressure is increased. Between points B and C on the intrusion curve in Fig. 11.1, intrusion commences into a new range of cavities which, if cylindrical in shape, would possess circular openings of about 100 Å radius. This range of intrusion occurs at a pressure substantially greater than was

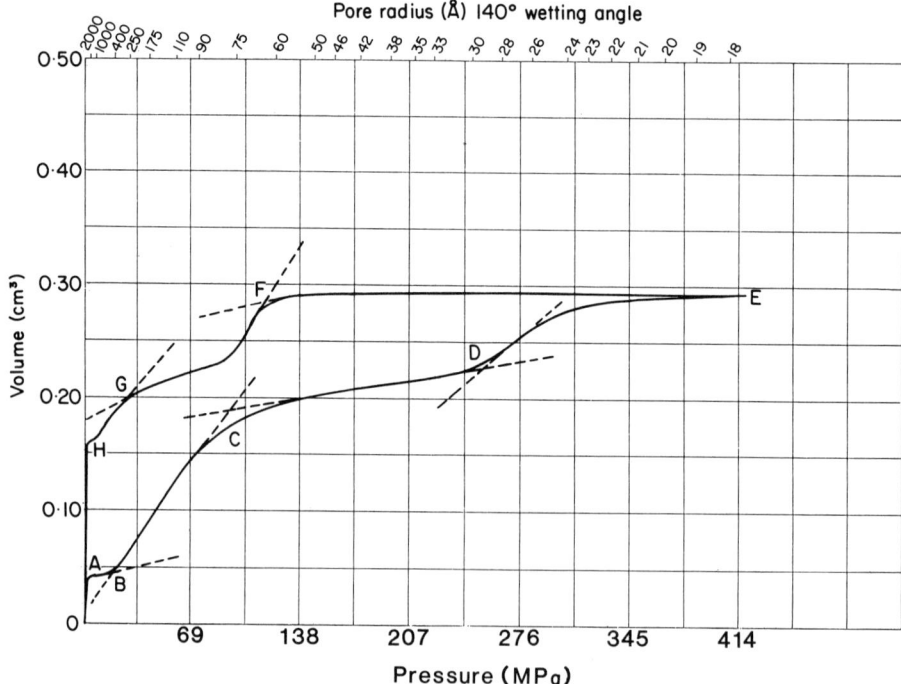

Figure 11.1 Mercury porosimetry data for an alumina sample.

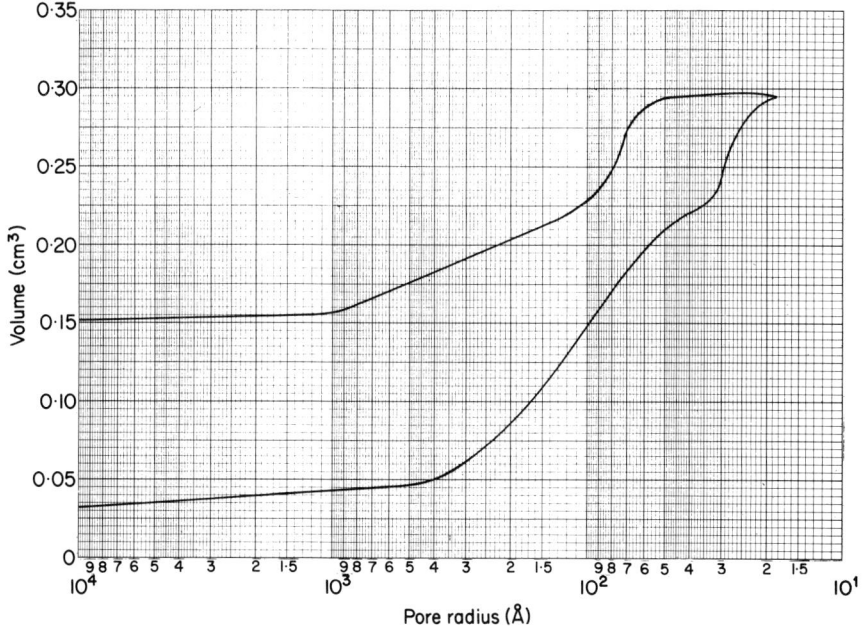

Figure 11.2 Plot of volume versus log radius.

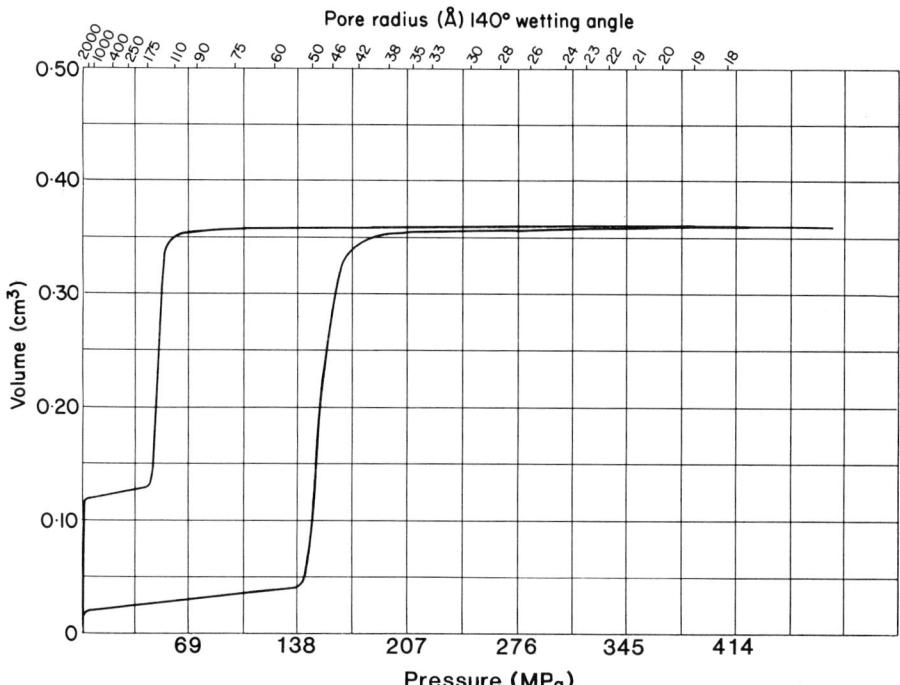

Figure 11.3 Intrusion–extrusion curves of porous glass.

102 Interpretation of mercury porosimetry data

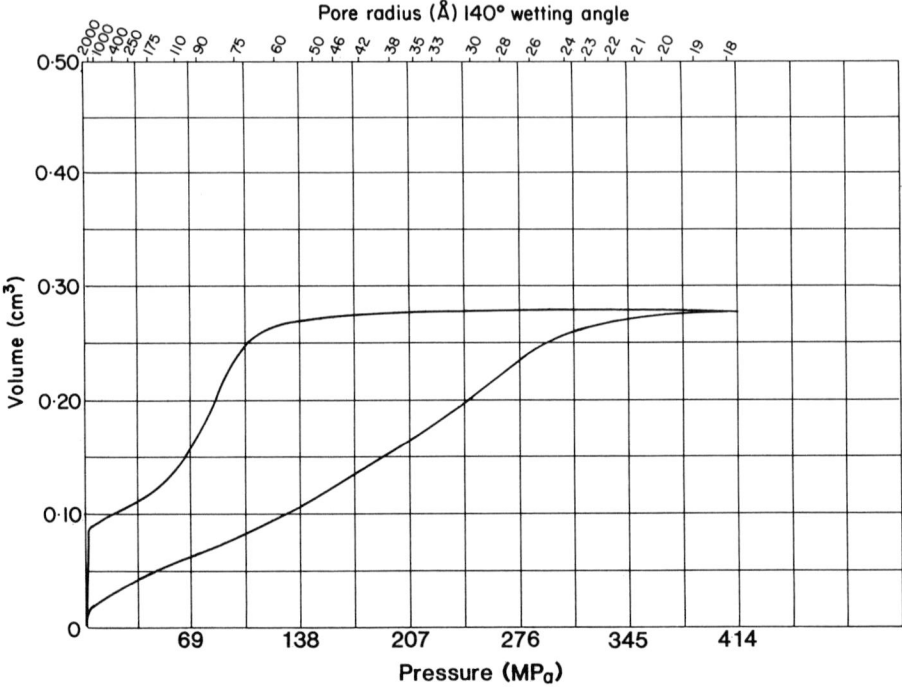

Figure 11.4 Intrusion–extrusion curves of an alumina sample.

required to fill the interparticle voids. Therefore, mercury is intruding into small pores within the particles. At point D, intrusion commences into a range of pores with even smaller radii (29 Å) after which no further intrusion takes place up to the maximum pressure corresponding to point E on the curve. The slight slope between points C and D indicates a small amount of porosity in the radius range of about 30 to 75 Å.

During depressurization, pores which commenced filling at point D begin to empty at point F and at lower pressures pores that filled between points B and C begin to empty at G. At point H on the extrusion curve, the cycle is terminated. The intrusion–extrusion cycle does not close when the initial pressure is reached, indicating that some mercury has been permanently entrapped by the sample. Often, at the completion of an intrusion–extrusion cycle mercury will slowly continue to extrude, sometimes for hours.

Figs. 11.1 and 11.2 illustrate mercury porosimetry data of a bimodal size distribution. However, other types of less typical curves are often encountered. For example, samples of controlled porous glass exhibit intrusion–extrusion curves illustrated by Fig. 11.3, in which all the pores are essentially of one radius.

On the other hand, some samples possess a wide and continuous range of pore sizes, resulting in the porosimetry curve shown in Fig. 11.4.

Common features of porosimetry curves 103

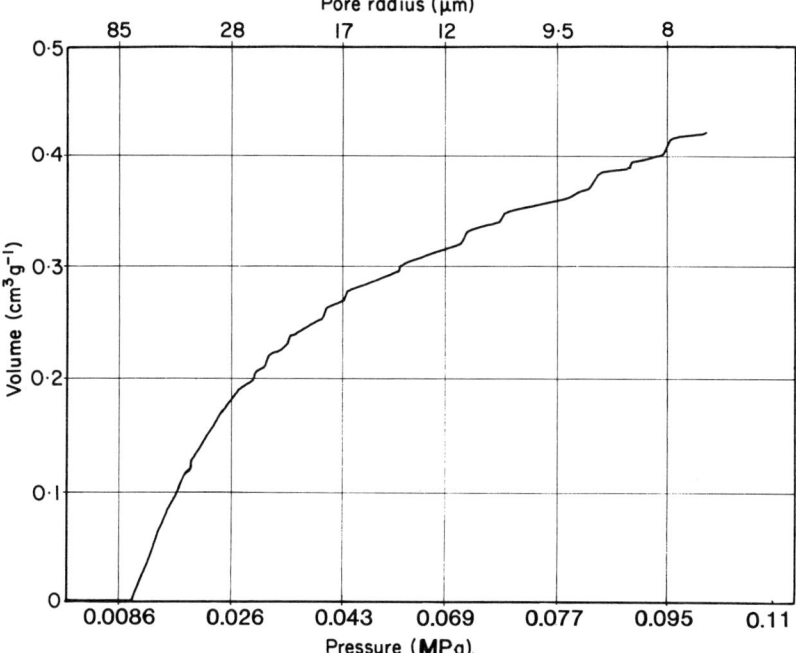

Figure 11.5 Low-pressure intrusion curve.

Still another type of porosimetry curve has recently [99] been reported. Fig. 11.5 illustrates mercury intrusion into distinct pore sizes in crystals of calcium hydrogen phosphate. The vertical steps on the curve indicate very narrow bands of pore radii between which no pores exist. Therefore, pores do not always exist over a continuum of radii.

The authors have found similar stepwise intrusion on other materials. The low pressure (0.003–0.1 MPa) intrusion curve in Fig. 11.5 was obtained using a scanning porosimeter [100] which continuously plots the pressure and corresponding intruded volume on an XY recorder. Only in this continuous manner can the exact position and height of each intrusion step be fully determined.

11.3 COMMON FEATURES OF POROSIMETRY CURVES

All mercury intrusion–extrusion curves have certain characteristic in common. These include:

1. On porous powdered samples, intrusion takes place at low pressures as mercury penetrates the large interparticle voids. Additional intrusion occurs at higher pressures into pores within the particles.

2. All porosimetry curves exhibit hysteresis, that is, the path followed by the extrusion curve is not the same as the intrusion path. At a given pressure, the volume indicated on the extrusion curve is greater than that on the intrusion curve and for a given volume the pressure indicated on the intrusion curve is greater than that on the extrusion curve.
3. Upon completion of a first intrusion–extrusion cycle, some mercury is always retained by the sample, thereby preventing the loop from closing.
4. Intrusion–extrusion cycles after the first will continue showing hysteresis but eventually the loop will close, showing that entrapment of mercury eventually ceases. On most samples, the loop closes after the second pressurization–depressurization cycle.
5. Each step on an intrusion curve has a corresponding step on the extrusion curve at a lower pressure.

11.4 SOLID COMPRESSIBILITY

The compressibility β of solids is the fractional change in volume per unit pressure change and is expressed as

$$\beta = \frac{1}{V}\left(\frac{dV}{dP}\right) \qquad (11.3)$$

The value of β for most solids lies in the range 10^{-11} to 10^{-12} (dynes/cm^2)$^{-1}$ or about 10^{-3} to 10^{-4} (Pa)$^{-1}$. This implies that a sample with a volume of about 1 cm^3 will be compressed by one-millionth to one ten-millionth of a cubic centimeter for each pound of applied pressure. Therefore, at pressures as high as 414 MPa, the sample will compress by only 0.006 to 0.06 cm^3. However, some softer plastics and other materials may have greater compressibilities for which corrections should be made. Some materials have occluded pores without access to the surface and can collapse on pressurization. Fig. 11.6 illustrates an intrusion–extrusion curve accompanied by compression and decompression curves for a sample of porous material.

The line from the origin to point A is the compressibility curve. It is from this line and not from the abscissa that intruded volumes are measured. Similarly, the line from point B to the ordinate is the decompression curve from which the extruded volumes are measured. The assumptions made when using this method for correcting intrusion–extrusion curves for compressibilities are:

1. that the material is elastic.
2. that occluded pores or bubbles in the sample are also elastic and restore to their initial condition after decompression.
3. that the compressibility is constant over the entire pressure range. In fact, solid compressibilities are slightly nonlinear, decreasing with increasing pressure.

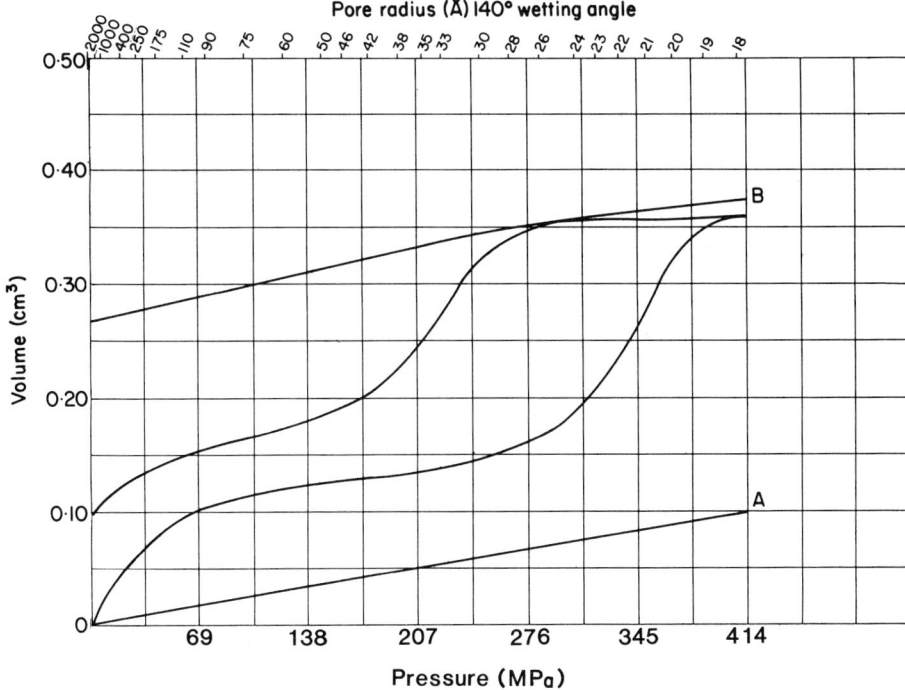

Figure 11.6 Porosimetry and compression curves.

11.5 SURFACE AREA FROM INTRUSION CURVES

The surface area of all pores and voids filled up to pressure P can be obtained from mercury intrusion data.

Fig. 11.7 is a cumulative pore volume intrusion curve which shows the summation of volume intruded into the pores and interparticle voids plotted versus the applied pressure. Recognizing that the increase in interfacial area ΔA from equation (10.22) is effectively the pore and void surface area S, this equation can be rewritten as

$$S\gamma|\cos\theta| = P\Delta V \tag{cf.10.22}$$

Then, pores in the radius range dr which take up volume dV will possess area dS as given by

$$dS = \frac{PdV}{\gamma|\cos\theta|} \tag{11.4}$$

106 Interpretation of mercury porosimetry data

The area of pores filled up to radius r is given by integration of the preceding equation which is, assuming constancy of surface tension and wetting angle, given by

$$S = \frac{1}{\gamma |\cos\theta|} \int_0^V P dV \qquad (11.5)$$

The surface area calculated by the above method is modelless and assumes no specific pore geometry.

For various pressure units with V in cubic centimeters, $\theta = 140°$, and $\gamma = 480$ erg cm^{-2}, equation (11.5) becomes

$$S = 2.72 \int_0^V PdV \text{ (m}^2\text{)} \quad P \text{ in MPa} \qquad (11.6)$$

$$S = 2.645 \times 10^{-5} \int_0^V PdV \text{ (m}^2\text{)} \quad P \text{ in kg m}^{-2} \qquad (11.7)$$

By graphical integration, using the cumulative intrusion curve in Fig. 11.7, the surface area of all pores filled by mercury up to 414 MPa ($r = 0.00178\,\mu$m) is calculated as follows:

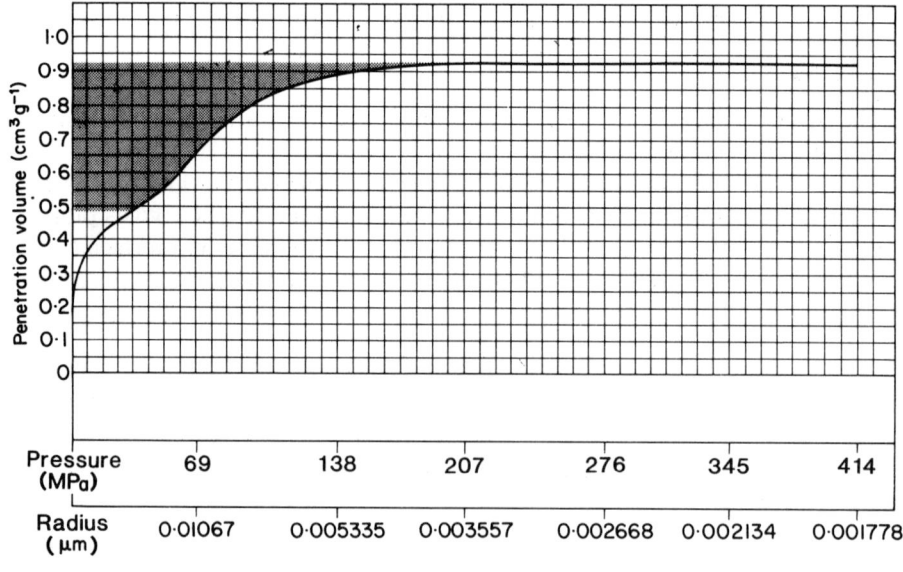

Figure 11.7 Cumulative pore volume plot.

1. The area above the total curve to the maximum intruded volume, measured by planimetry, is 55.2 cm².

2. Normalization factor $= \dfrac{69 \text{ MPa}(10^{-1}\text{cm}^3\text{g}^{-1})}{4\text{ cm}^2 \text{ of graph paper}}$

$$= \dfrac{1.725 \text{ MPa cm}^3\text{g}^{-1}}{\text{cm}^2} \qquad (11.8)$$

3. $\int_0^V PdV = (55.2\text{ cm}^2)\dfrac{1.725 \text{ MPa cm}^3\text{g}^{-1}}{\text{cm}^2}$

$$= 95.22 \text{ MPa cm}^3\text{g}^{-1} \qquad (11.9)$$

4. $S = 2.72 \int_0^V PdV = (2.72)(95.22)$

$$= 259 \text{ m}^2\text{g}^{-1} \qquad (11.10)$$

The surface area calculated in equation (11.10) represents the total area of all pores between ambient pressure ($r = 7.25$ μm) and 414 MPa ($r = 0.00178$ μm).

The surface area of pores in any selected radius interval can also be obtained by the preceding method. For example, the graphical area over the curve between 34.5 MPa ($r = 0.0213$ μm) and 414 MPa ($r = 0.00178$ μm), indicated by the shaded area in Fig. 11.7, is 50.1 cm². Then,

$$S = 2.72 \times 50.1 \times 1.725 = 235 \text{ m}^2\text{g}^{-1} \qquad (11.11)$$

Therefore, the surface area of pores in the range 0.0213–0.00178 μm is 235 m² g⁻¹. The area of the large pores and interparticle voids is given by the difference between the total surface area and the area obtained in equation (11.11) or 24 m² g⁻¹.

Some newly-developed automated porosimeters have associated computer-aided data reduction capabilities which perform the above integration as rapidly as data is acquired [100].

11.6 PORE SIZE DISTRIBUTION

When the radius of a cylindrical pore is changed from r to $r - dr$ the corresponding decremental change in the pore volume V is

$$dV = -2n\pi r \ell dr \qquad (11.12)$$

where n is the number of pores with radius r and length ℓ. However, when pores are filled according to the Washburn equation, the volumetric change with decreasing radius does not necessarily decrease since it corresponds to the filling of a new group of pores. Thus, when the pore radius into which intrusion occurs changes from r to $r - dr$ the corresponding volume change is given by

$$dV = -D_v(r)dr \qquad (11.13)$$

where $D_v(r)$ is the volume pore size distribution function, defined as the pore volume per unit interval of pore radius.

Differentiation of the Washburn equation (10.24), assuming constancy of γ and θ, yields

$$P\,dr + r\,dP = 0 \qquad (11.14)$$

Combining equations (11.13) and (11.14) gives

$$dV = D_v(r)\frac{r}{P}dP \qquad (11.15)$$

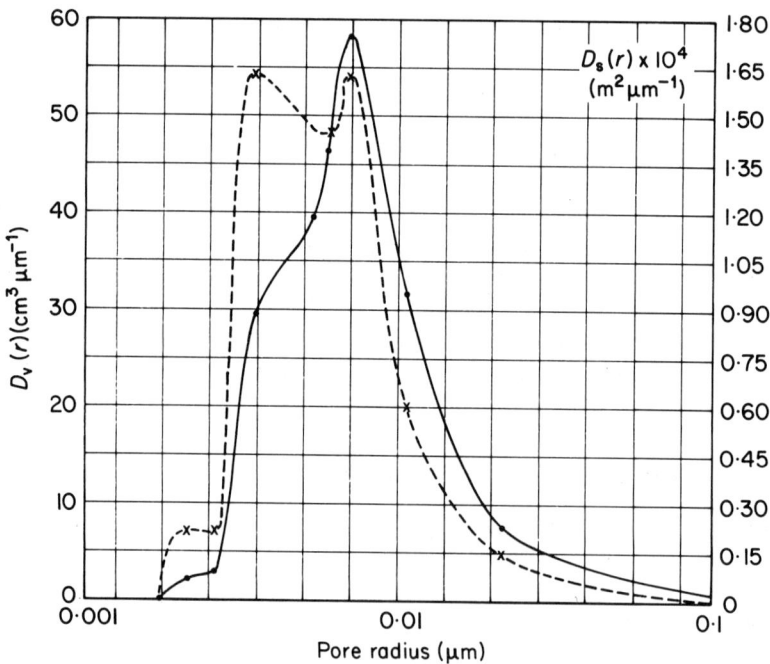

Figure 11.8 Distribution curves for volume $D_v(r)$, ($\bullet$) and area $D_s(r)$, ($\times$).

Rearranging,

$$D_v(r) = \frac{P}{r}\left(\frac{dV}{dP}\right) \qquad (11.16)$$

Equation (11.16) represents a convenient means of reducing the cumulative curve to the distribution curve which gives the pore volume per unit radius interval.

Fig. 11.8 is the distribution prepared from Fig. 11.7. A series of values of the ratio $\Delta V/\Delta P$ is taken from Fig. 11.7 or preferably from the raw data.

Each value of $\Delta V/\Delta P$ is multiplied by the pressure at the upper end of the interval and divided by the corresponding pore radius. Alternatively, the mean pressure and radius in an interval may be used to calculate $D_v(r)$. The resulting $D_v(r)$ values are plotted versus the pore radius. Table 11.1 contains the raw data from which the cumulative curve, Fig. 11.7, is obtained, as well as the tabulated calculations necessary to prepare the distribution curve, Fig. 11.8.

Fig. 11.8 is plotted semilogarithmically to include the wide range of pore radii. The function $D_v(r)$, as is true for surface area and all subsequently described functions, is more conveniently obtained by computer-aided data reduction, as mentioned previously.

Unlike surface area calculations, the volume distribution function and all subsequently discussed functions are based on the model of cylindrical pore geometry.

Table 11.1 Data and calculations for the cumulative (Fig. 11.7) and distribution (Fig. 11.8) curves

P (MPa)	ΔP (MPa)	$r \times 10^3$ (μm)	V (cm^3)	ΔV (cm^3)	$\Delta V/\Delta P \times 10^3$ (cm^3 MPa^{-1})	$D_v(r)$ (cm^3 μm^{-1})	$D_s(r) \times 10^3$ (m^2 μm^{-1})
0.345	0.345	213.0	0.085	0.085	0.246	0.039	0.000
1.38	1.035	533.0	0.142	0.057	0.0551	0.142	0.000
3.45	2.070	213.0	0.251	0.109	0.0527	0.852	0.001
6.90	3.45	106.0	0.330	0.079	0.0229	1.482	0.003
34.5	27.6	21.3	0.461	0.131	0.0047	7.602	0.071
69.0	34.5	10.6	0.633	0.172	0.0050	32.364	0.605
103.5	34.5	7.11	0.772	0.139	0.0040	58.228	1.638
138.0	34.5	5.33	0.825	0.053	0.0015	38.837	1.457
207.0	69.0	3.5	0.860	0.035	0.0005	29.073	1.633
276.0	69.0	2.6	0.862	0.002	0.00003	3.101	0.232
345.0	69.0	2.13	0.863	0.001	0.00001	1.620	0.152
414.0	69.0	1.7	0.863	0.000			

11.7 VOLUME LN RADIUS DISTRIBUTION FUNCTION

Another useful function sometimes used in place of the volume distribution function is $D_v(\ln r)$ which can be expressed as

$$D_v(\ln r) = \frac{dV}{d\ln r} = r\left(\frac{dV}{dr}\right) = rD_v(r) \tag{11.17}$$

From equation (11.16) one obtains

$$D_v(\ln r) = P\left(\frac{dV}{dP}\right) = \frac{dV}{d\ln P} \tag{11.18}$$

The volume ln radius distribution function serves to reduce the wide range of values which the volume distribution function can exhibit. For example, if the volume distribution function ratio for intrusion into pores of 1000 Å and 50 Å is 10^6 then the ratio of the volume ln radius functions would be

$$\frac{D_v(\ln r) \text{ at } 50 \text{ Å}}{D_v(\ln r) \text{ at } 1000 \text{ Å}} = \frac{50}{1000} \times 10^6 = 5 \times 10^4 \tag{11.19}$$

11.8 PORE SURFACE AREA DISTRIBUTION

The pore surface distribution $D_s(r)$ is the surface area per unit pore radius. By chain differentials, one can write

$$D_s(r) = \left(\frac{dS}{dV}\right)\left(\frac{dV}{dr}\right) \tag{11.20}$$

Assuming cylindrical pore geometry, then

$$\frac{dS}{dV} = \frac{2}{r} \tag{11.21}$$

and

$$D_s(r) = \frac{2}{r} D_v(r) \tag{11.22}$$

The last column in Table 11.1 gives values of $D_s(r)$ calculated from equation (11.22). $D_s(r)$ is also plotted in Fig. 11.8.

The total area under the volume and area distribution curves is proportional to the total pore volume and pore area, respectively. By taking the ratio of the graphical area A_g in any interval to the total graph area A_t the pore volume or surface area in any interval can be calculated

$$V_{\text{interval}} = V_t \frac{A_g}{A_t} \tag{11.23}$$

$$S_{\text{interval}} = S_t \frac{A_g}{A_t} \tag{11.24}$$

11.9 PORE LENGTH DISTRIBUTION

The volume distribution function $D_v(r)$ represents the volumetric uptake in a unit interval of pore radii, irrespective of the variation in the number or the length of the pores. When $D_v(r)$ is divided by πr^2, the mean cross-sectional area in the radius interval, the new function reflects only variations in the pore length and is called the length distribution function, $D_\ell(r)$. Thus

$$D_\ell(r) = \frac{D_v(r)}{\pi r^2} \tag{11.25}$$

This function effectively assigns the variation in the volumetric uptake in a radius interval to differences in the pore lengths and assumes constancy of pore population in each interval.

11.10 PORE POPULATION

When a volume ΔV is intruded into a narrow pore radius range of Δr, centered about a unit radius r, one can write

$$\frac{\Delta V}{\pi r^2} = n\ell \tag{11.26}$$

where n is the number of pores of average length, ℓ. Since, for sufficiently small values of Δr

$$\frac{\Delta V}{\Delta r} = D_v(r) \tag{11.27}$$

it follows that

112 Interpretation of mercury porosimetry data

$$\frac{D_v(r)\Delta r}{\pi r^2} = n\ell \tag{11.28}$$

For equal small pore radius intervals, assuming equal pore lengths, equation (11.28) can be used to obtain relative pore populations, viz.,

$$\frac{\dfrac{D_v(r_1)}{r_1^2}}{\dfrac{D_v(r_2)}{r_2^2}} = \frac{n_1}{n_2} \tag{11.29}$$

Since

$$D_v(r) = \frac{2D_s(r)}{r} \tag{11.30}$$

it follows that

$$\frac{n_1}{n_2} = \frac{\dfrac{D_v(r_1)}{r_1^2}}{\dfrac{D_v(r_2)}{r_2^2}} = \frac{\dfrac{D_s(r_1)}{r_1}}{\dfrac{D_s(r_2)}{r_2}}$$

In this manner it is possible to obtain relative pore populations throughout the entire range of measurable pores.

11.11 PLOTS OF POROSIMETRY FUNCTIONS

Figs. 11.9 to 11.14 illustrate some of the previously discussed functions plotted linearly versus pressure (Figs. 11.9(a) to 11.14(a)) and logarithmically versus radius (Figs. 11.9(b) to 11.14(b)), such that each decade of pore radius assumes a uniform interval. A mixture of porous glasses was used as the sample and the data was obtained on a Quantachrome Autoscan-60 porosimeter.

Figs. 11.10(a) and 11.10(b) are normalized such that the maximum intruded volume from Figs. 11.9(a) and 11.9(b) are adjusted to 100% or full scale on the volume axis. These plots are, therefore, labelled % V on the volume axis since any point on that axis becomes the percentage of the maximum intruded volume.

For plots of $D_v(r)$, $D_s(r)$ and S, the height of the curve at any point along the pressure or radius axis reflects not only the intruded volume at that point but also the fact that the numerical value of the function is determined by the

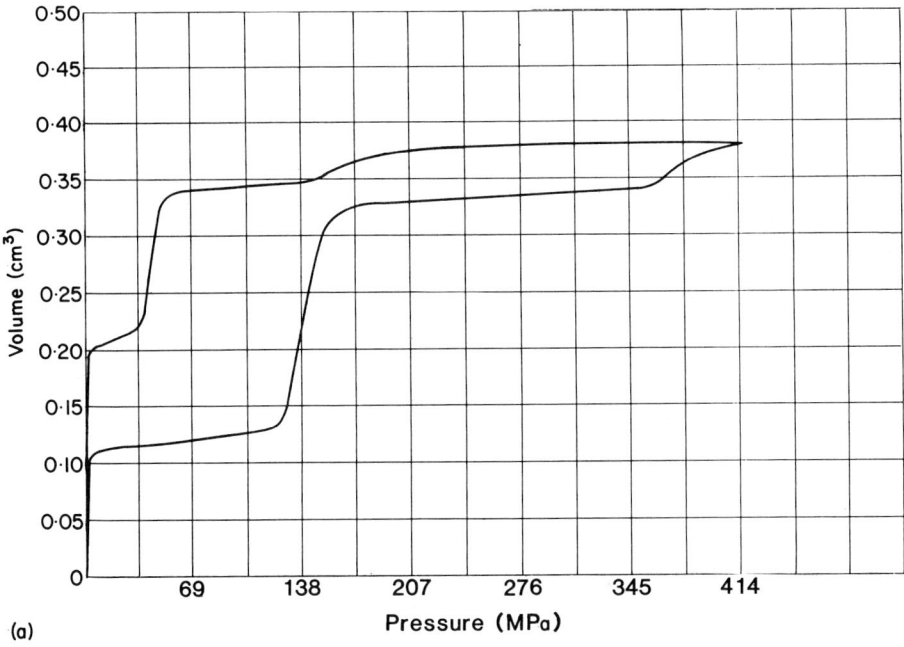

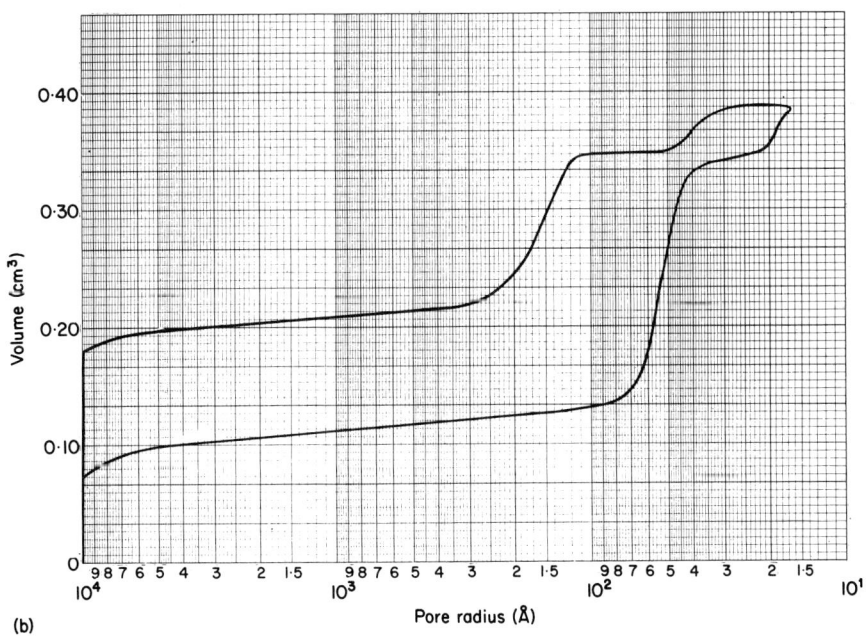

Figure 11.9 (a) Volume versus pressure plot, (b) Volume versus radius plot.

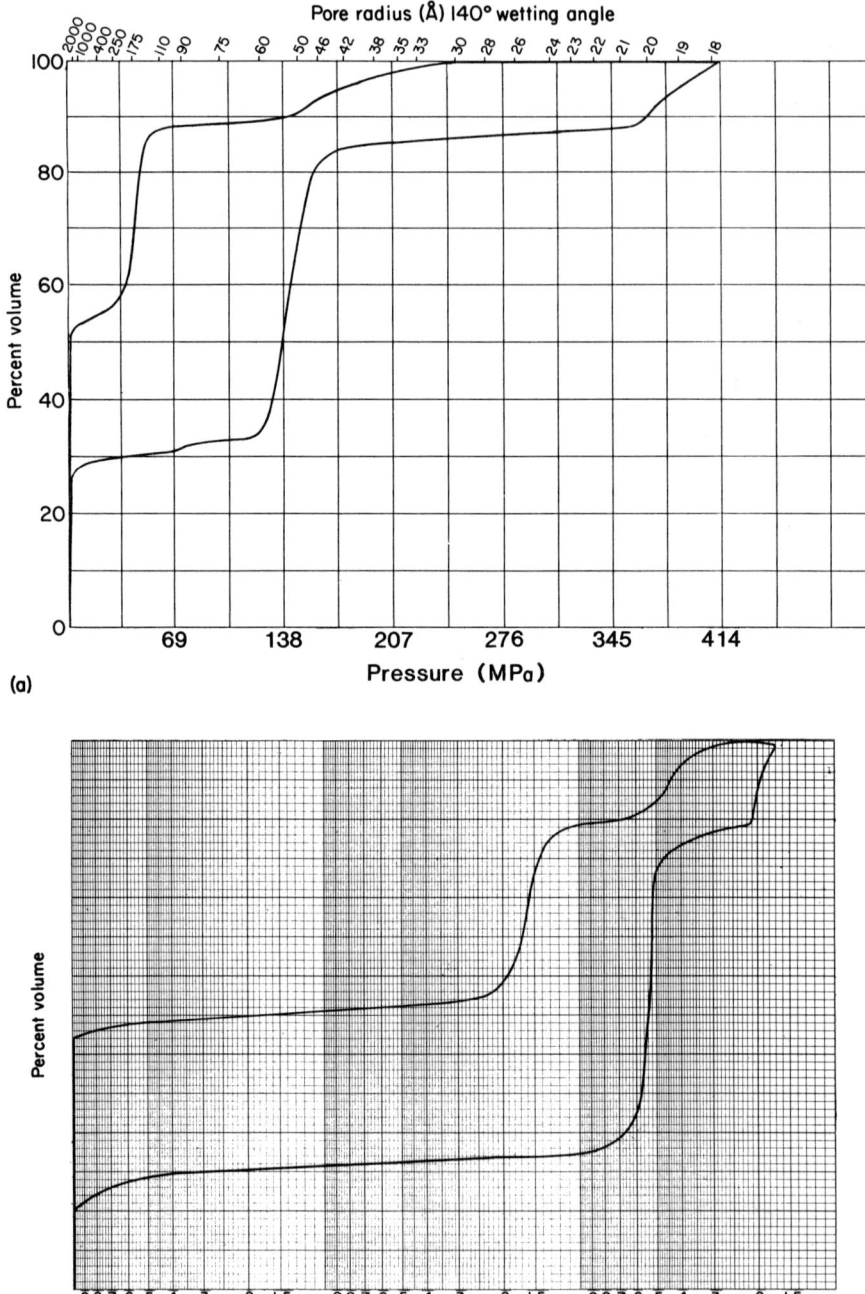

Figure 11.10 (a) Percentage volume versus pressure plot, (b) Percentage volume versus radius plot.

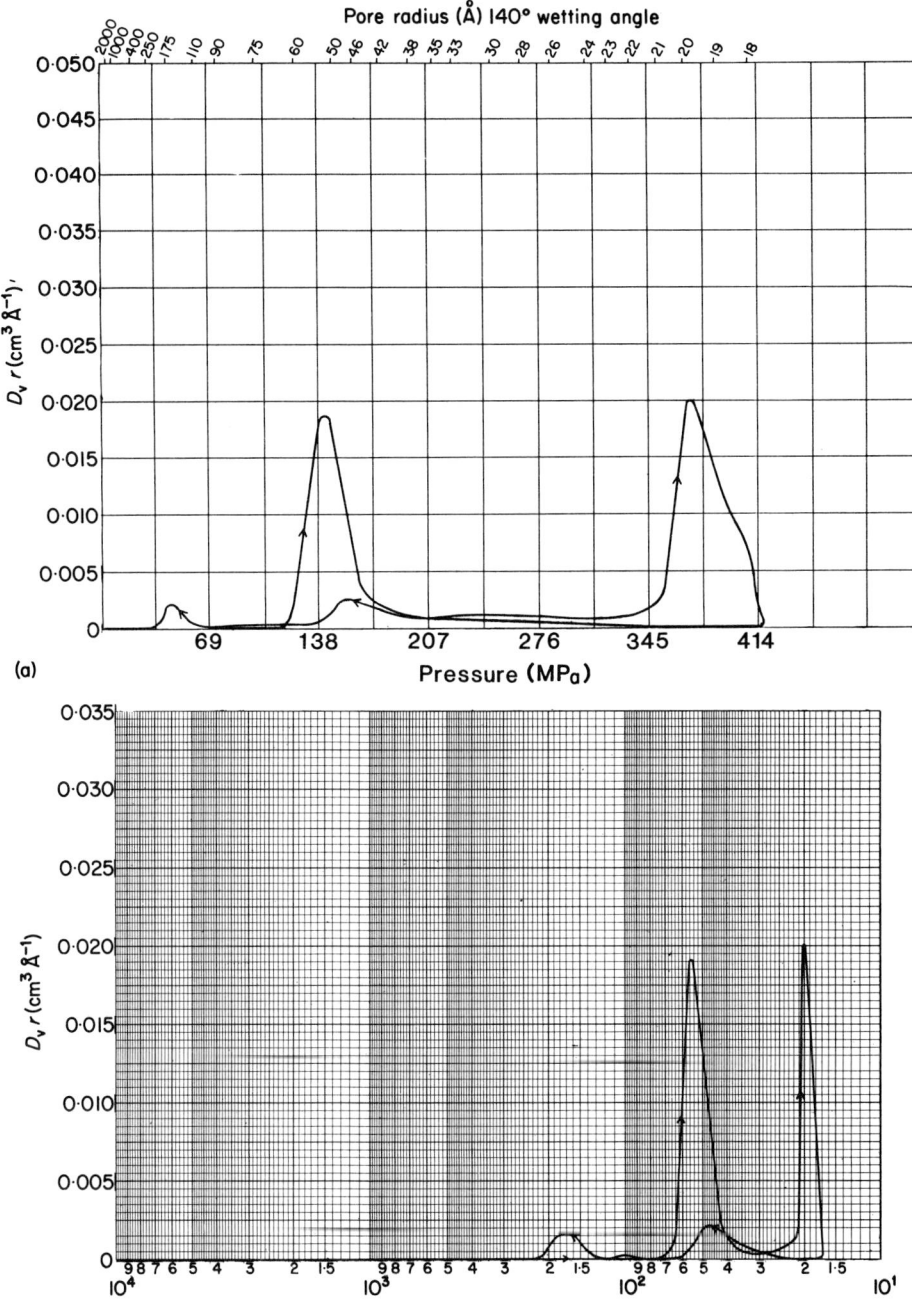

Figure 11.11 (a) Pore volume distribution versus pressure plot. Intrusion curve (→); extrusion curve (←), (b) Pore volume distribution versus radius plot. Intrusion curve (→); extrusion curve (←).

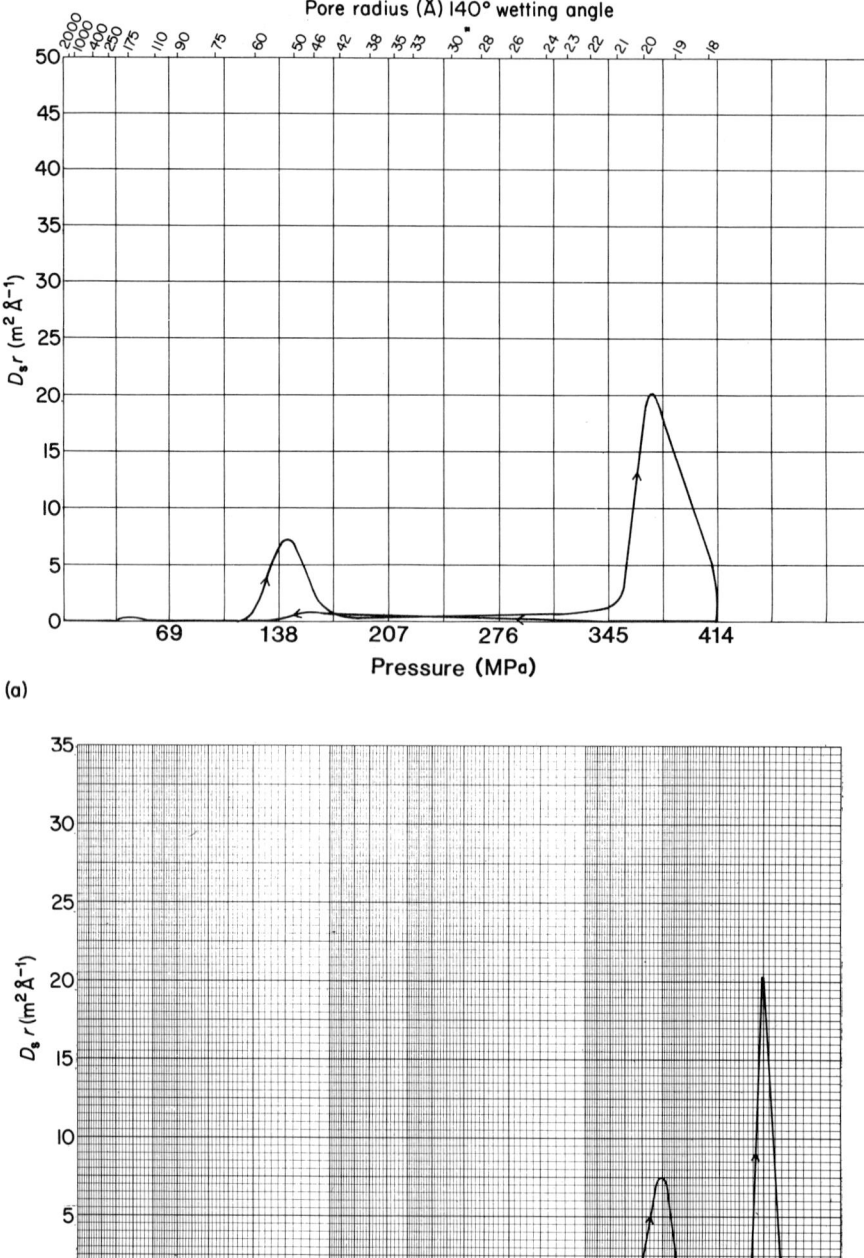

Figure 11.12 (a) Pore surface area distribution versus pressure plot. Intrusion curve (→); extrusion curve (←), (b) Pore surface area distribution versus radius plot. Intrusion curve (→); extrusion curve (←).

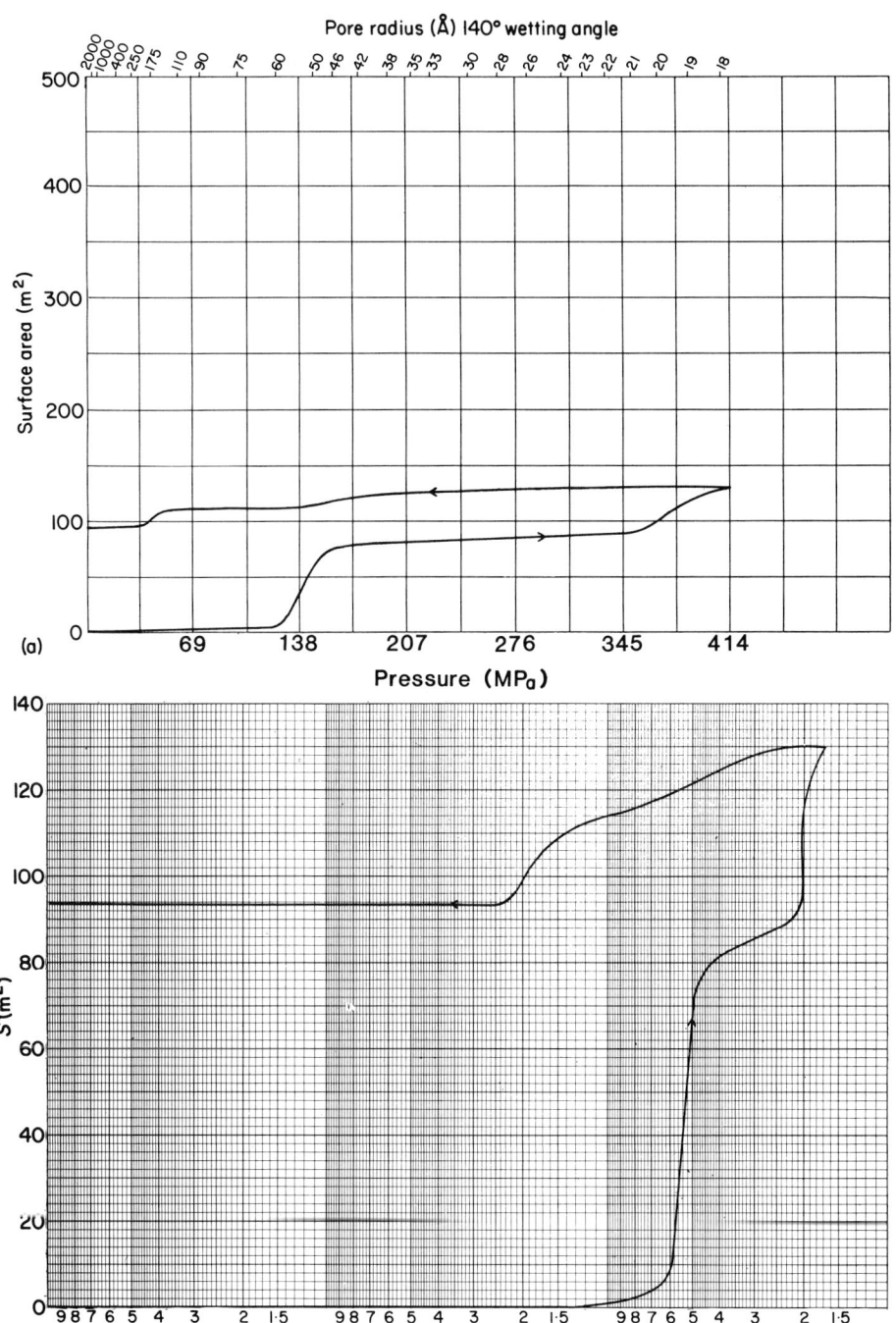

Figure 11.13 (a) Surface area versus pressure plot. Intrusion curve (→); extrusion curve (←), (b) Surface area versus radius plot. Intrusion curve (→); extrusion curve (←).

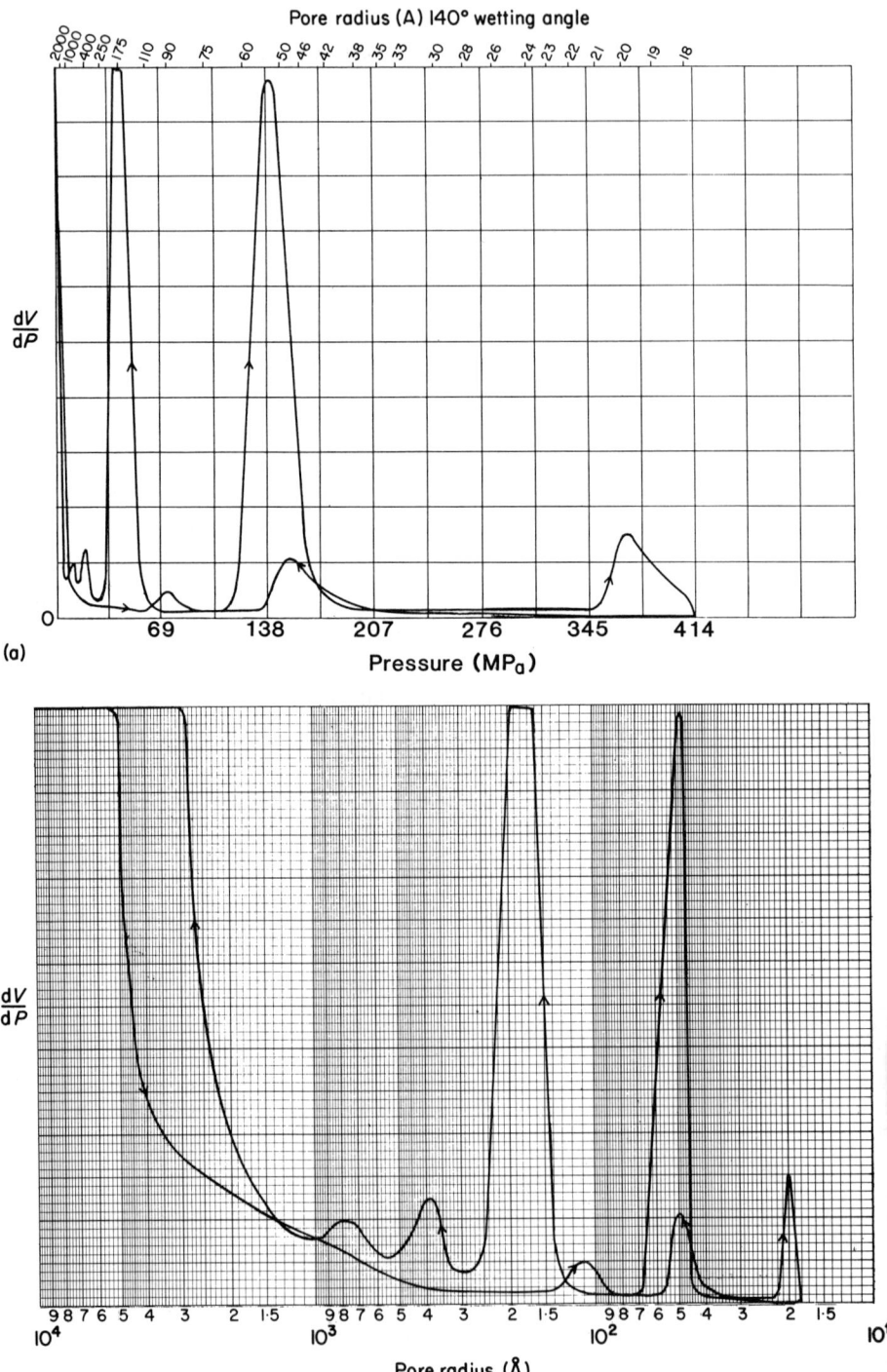

Figure 11.14 (a) Derivative plot: dV/dP versus pressure. Intrusion curve ($\rightarrow$); extrusion curve ($\leftarrow$), (b) Derivative plot: dV/dP versus radius. Intrusion curve ($\rightarrow$); extrusion curve ($\leftarrow$).

pressure at which a given volume is intruded as shown, for example, by equations (11.5), (11.16) and (11.22). The $D_v(r)$, $D_s(r)$ and S plots are illustrated in Figs. 11.11, 11.12, and 11.13, respectively. A small quantity of intrusion or extrusion at high pressure will often generate a function which is numerically greater than that caused by the same or an even larger amount at a lower pressure.

Plots of the derivative of the cumulative curves, dV/dP versus pressure or radius, are often useful for the determination of the radius or pressure at which the maximum volume intrudes or extrudes. Figs. 11.14(a) and 11.14(b) are derivative plots calculated from the cumulative curves in Figs. 11.9(a) and 11.9(b).

The functions $D_v(r)$ and $D_s(r)$ show a smooth curve at lower pressures and become more irregular at higher pressures. The mathematical nature of these functions requires that the plots become more sensitive to slope changes as the pressure increases. The irregularities, small peaks, at higher pressures reflect small amounts of intrusion into discontinuous ranges of pores. In order to resolve this fine structure it is necessary to obtain the data on a continuous basis with subsequent digital data reduction employing several hundred small increments.

11.12 COMPARISONS OF POROSIMETRY AND GAS ADSORPTION

The useful range of the Kelvin equation is limited at the narrow pore end by the question of its applicability (see Section 8.5) and at the wide pore end, measurements are limited by the rapid cha.ige of the center core radius with relative pressure.

Mercury porosimetry has somewhat the same constraints at the narrow pore end of its range, in that the same questions arise regarding the constancy of surface tension and wetting angle for mercury as exist for an adsorbate. Consequently, both methods have nearly the same lower limit which is about 18 Å pore radius for mercury intrusion at 414 MPa. However, at the wide pore end porosimetry does not have the limitation of the Kelvin equation and for example, at 0.0069 MPa pore volumes can be measured in pores of 107 micrometer radius or 1.07×10^6 Å.

Comparisons have been made in the range where the two methods overlap and they generally show reasonable agreement. Zwietering [101] obtained distribution curves from mercury porosimetry and the nitrogen isotherms on chromium oxide–iron oxide catalysts. Each curve showed a narrow distribution with the peak near 150 Å, the nitrogen curve being slightly narrower and higher. Using nitrogen, Joyner, Barrett, and Skold [102] obtained good agreement between the two methods. By adjusting the wetting angle for mercury on charcoals to values between 130° to 140°, they were able closely to match the curve produced from the nitrogen isotherm. Cochran and Cosgrove

[103], using *n*-butane, reported total pore volumes of $0.458\,\text{cm}^3\,\text{g}^{-1}$ and $0.373\,\text{cm}^3\,\text{g}^{-1}$ with adsorption and porosimetry, respectively. The maximum pore radius was under $500\,\text{Å}$ and they attributed the difference to *n*-butane entering pores narrower than $30\,\text{Å}$. Dubinin *et al.* [104] found that nitrogen and benzene gave Type I isotherms with hysteresis on several activated carbons. The pore distributions based on the Kelvin equation agreed with porosimetry measurements.

Comparisons have been made between surface areas measured by porosimetry and gas adsorption [105] as well as by permeability [106] with results ranging from excellent to poor.

The most definitive surface area measurements are probably those made by nitrogen adsorption using the BET theory. Neither the Brunauer, Emmett, and Teller (BET) theory nor equation (11.5), used to calculate surface area from mercury intrusion data, makes any assumptions regarding pore shape for surface area determinations. When these two methods are compared, there is often surprisingly good agreement. When the two methods do not agree, it does not imply the theoretical failure of either one. Indeed, the differences between the two methods can be used to deduce meaningful information which neither method alone can supply. For example, when the BET data is large compared to the area measured by porosimetry, the implication is that there is a substantial volume of pores smaller than those penetrated by mercury at the maximum pressure.

If the porosimeter can generate $414\,\text{MPa}$ of hydraulic pressure, the minimum radius into which intrusion can occur will be about $18\,\text{Å}$. Assuming that pores centered about $15\,\text{Å}$ radius are present and have a volume of $0.01\,\text{cm}^3$, an approximation of their surface area can be made by assuming cylindrical geometry. Thus,

$$S = \frac{2V}{r} = \frac{2 \times 0.01}{15 \times 10^{-8}} \times 10^{-4} = 13.3\,\text{m}^2 \tag{11.31}$$

Therefore, the area measured by porosimetry would be approximately $13.3\,\text{m}^2$ less than that measured by the BET method.

Another factor which can lead to BET areas slightly higher than those from porosimetry is pore wall roughness. Slight surface roughness will not alter the porosimetry surface area since it is calculated from the pore volume while the same roughness will be measured by gas adsorption.

Cases that lead to porosimetry-measured surface areas exceeding those from nitrogen adsorption can result from ink-bottle shaped pores having a narrow entrance with a wide inner body. Intrusion into the wide inner body will not occur until sufficient pressure is applied to force the mercury into the narrow entrance. It will, therefore, appear as if a large volume intruded into narrow pores, generating an excessively high calculated surface area.

12
Hysteresis, entrapment and contact angle

12.1 INTRODUCTION

Cumulative volume curves generated by intruding mercury into porous samples are not followed as the pressure is lowered and mercury extrudes out of the pores. In all cases, the depressurization curve lies above the pressurization curve and the hysteresis loop does not close even when the pressure is returned to zero, indicating that some mercury is entrapped in the pores. Usually after the sample has been subjected to a first pressurization–depressurization cycle, no additional entrapment occurs during subsequent cycles. In some cases, however, a third or even fourth cycle is required before entrapment ceases.

The pressure–volume (P–V) work associated with an initial intrusion–extrusion cycle can be calculated from the areas above curves A and C to the maximum intruded volume, indicated by the horizontal dotted line in Fig. 12.1. (The work associated with any graphical area can be calculated with reference to equations (11.8) and (11.9)). This P–V work, $\oint dW_1$, can be expressed as

$$\oint dW_1 = \int_0^V P_i dV_i + \int_V^{V'} P_e dV_e > 0 \qquad (12.1)$$

where P_i and P_e are intrusion and extrusion pressures and V_i and V_e are intrusion and extrusion volumes, respectively. The limit V' is used in the extrusion integral in equation (12.1) to indicate that, at the completion of a cycle, some mercury has been retained by the sample.

Even in the absence of mercury entrapment, second or subsequent intrusion–extrusion cycles, curves B and C of Fig. 12.1, continue to show hysteresis with the P–V work, $\oint dW_2$, through a cycle given by

$$\oint dW_2 = \int_0^V P_i dV_i + \int_V^0 P_e dV_e > 0 \qquad (12.2)$$

122 Hysteresis, entrapment and contact angle

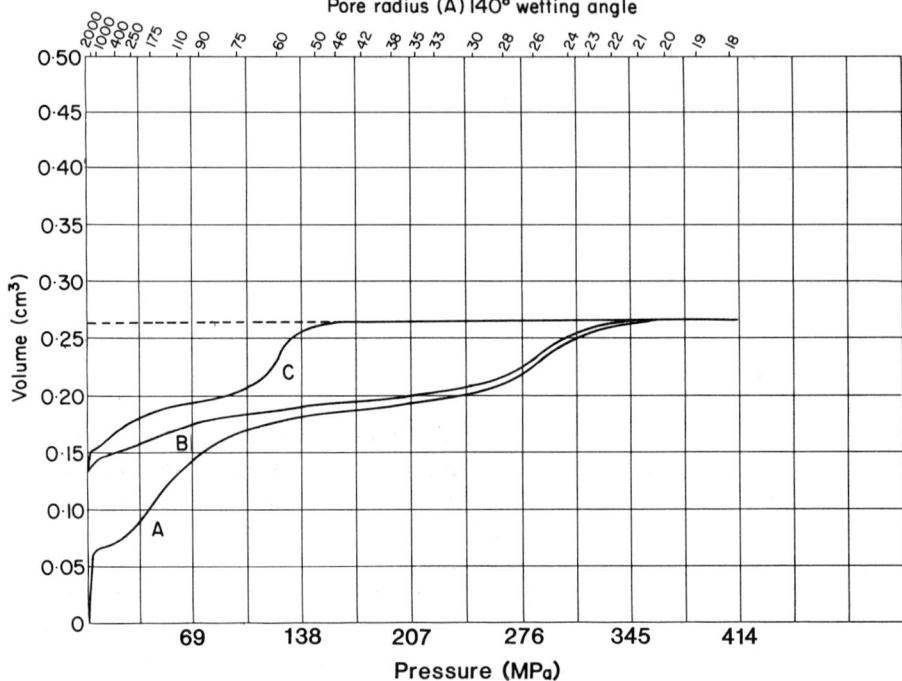

Figure 12.1 Cumulative volume versus pressure plot.

As illustrated in Fig. 12.1 the P–V work through a second cycle is always less than that through a first cycle since the area between curves B and C is less than that between A and C. Then,

$$\oint dW_1 > \oint dW_2 \tag{12.3}$$

The difference between the cyclic integrals $\oint dW_1$ and $\oint dW_2$ is the work associated with entrapment of mercury and can be evaluated from the area between curves A and B.

Equations (12.1) and (12.2) indicate that the work of intrusion always exceeds the work of extrusion. Therefore, the surroundings must experience a decrease in potential energy in order to provide the required work difference around an intrusion–extrusion cycle. After many cycles, the system must either store a boundless amount of energy or must convert the energy into heat as irreversible entropy production. Since, neither of these phenomena is observed, one must be able to show that around any intrusion–extrusion cycle, after entrapment has ceased, work is conserved, that is,

$$\oint dW = 0 \tag{12.4}$$

Inspection of a plot of volume versus radius for an intrusion–extrusion cycle (Fig. 12.1) would lead one to the conclusion, that at a given volume, mercury intrudes into pores of one size, but extrudes from pores of a smaller size. This anomaly and the question of energy conservation will be considered in subsequent sections of this Chapter.

12.2 CONTACT ANGLE CHANGES

Lowell and Shields [107, 108] have recently shown that superimposition of the intrusion and extrusion curves, when plotted as volume versus radius, can be achieved if the contact angle θ is adjusted from θ_i, the intrusion contact angle, to θ_e, the extrusion angle. Fig. 12.2 illustrates the resulting curves when the contact angles are properly adjusted. Curve A is the first intrusion curve using the intrusion contact angle and is identical to Curve A in Fig. 12.1. Curve B is the second intrusion curve using θ_i and is identical to curve B in Fig. 12.1. The difference between curves A and B in both figures along the volume axis reflects the quantity of entrapped mercury after completion of the first intrusion–extrusion cycle.

When the contact angle is changed from θ_i to the smaller value θ_e, the

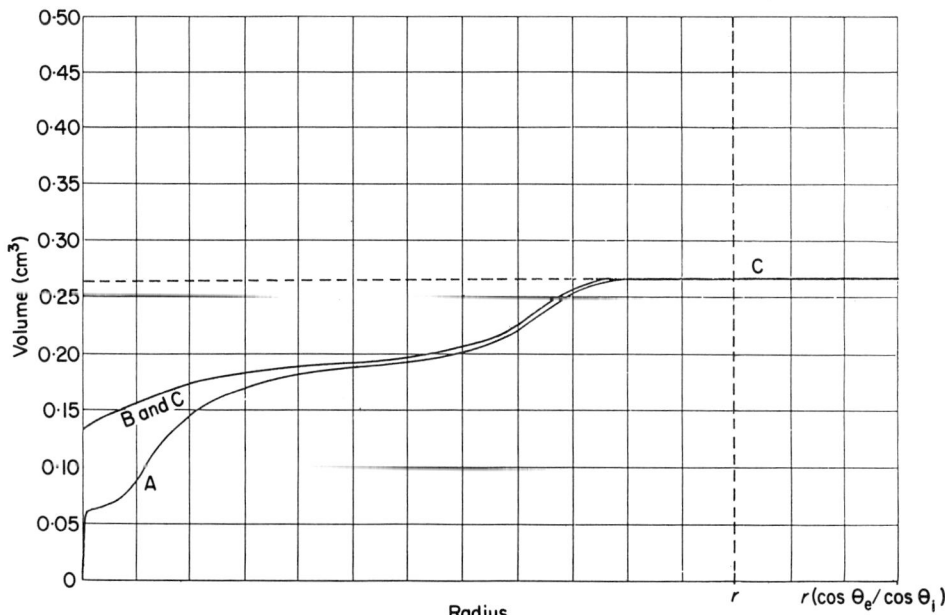

Figure 12.2 Cumulative volume versus radius plot.

pore radius calculated from the Washburn equation will decrease by the ratio of $r(\cos\theta_e/\cos\theta_i)$. This does not imply that intrusion occurs into pores of the smaller size since the actual intrusion contact angle is applicable. However, when the correction is made from θ_i to θ_e, at the start of extrusion, as shown in Fig. 12.2 as the linear region to the right of the vertical dotted line, the pore radius corresponding to the maximum pressure will be $r(\cos\theta_e/\cos\theta_i)$. Thus, when depressurization commences, shown in Fig. 12.2 as curve C, no extrusion occurs between $r(\cos\theta_e/\cos\theta_i)$ and r since no mercury has intruded into pores in this range. To the left of the dotted vertical line in Fig. 12.2, the extrusion curve C is superimposed on the second intrusion curve B, so that no hysteresis is exhibited. If mercury exhibits two or more different intrusion contact angles with a given material, superimposition of curves cannot be achieved.

12.3 POROSIMETRIC WORK

It has been demonstrated [108] that first and subsequent mercury intrusion–extrusion curves can be viewed as consisting of three significant areas when the extrusion contact angles are not corrected, as shown in Fig. 12.1.

The total area above the first intrusion curve, A in Fig. 12.1, to the maximum intruded volume indicated by the horizontal dotted line, corresponds to the P–V work of intrusion, W_i. This work term consists of three parts, the first of which is the work of entrapment, W_t, corresponding to the area between curves A and B. The second contribution to W_i is the work, $W_{\Delta\theta}$, associated with the contact angle change from θ_i to θ_e which corresponds to the area between curves B and C. The final contribution to the work of intrusion is the area between curve C and the maximum intruded volume which corresponds to the work of extrusion using the incorrect or intrusion contact angle. W_t, $W_{\Delta\theta}$ and W_e can be evaluated [108] either graphically from Fig. 12.1 or calculated from the equations below.

The work associated with entrapment of mercury is given by

$$W_t = \gamma|\cos\theta_i|A_t \tag{12.5}$$

where A_t is the area of mercury entrapped in all the pores.

The work associated with the contact angle change from θ_i to θ_e is

$$W_{\Delta\theta} = \gamma(|\cos\theta_i| - |\cos\theta_e|)A_{\Delta\theta} \tag{12.6}$$

in which $A_{\Delta\theta}$ is the area of mercury undergoing a change in contact angle and is the same as A_e, the area of mercury which extrudes from the pores.

The work of extrusion using the incorrect or intrusion contact angle is given by

$$W'_e = \gamma|\cos\theta_i|A'_e \tag{12.7}$$

When the correct extrusion contact angle is employed equation (12.7) becomes

$$W_e = \gamma|\cos\theta_e|A_e \tag{12.8}$$

It would appear from equations (12.7) and (12.8) that there are two different values for W_e, the work of extrusion. This is not the case. When equation (12.7) is used to calculate the work of extrusion, A_e is determined from Fig. 12.1 where the intrusion contact angle θ_i was used to obtain the extrusion curve. In fact, A'_e is a smaller area than that of the pores actually emptied. When equation (12.8) is employed, A_e represents the correct pore area and is larger than A'_e by the factor $\cos\theta_i/\cos\theta_e$. Thus, regardless of whether equation (12.7) or (12.8) is used, the calculated value of W_e will be the same.

Equations (12.5) through (12.8) assume no specific pore shape. By assuming cylindrical pore geometry the validity of these equations can be established. For a cylindrical pore the area in any pore interval is given by

$$A = \frac{2}{\bar{r}}V \tag{12.9}$$

where $\bar{r}$ is the mean pore radius in a narrow pore interval and V is the pore volume in that radius interval. Using equation (12.9) and expressing V as the volumetric difference between the appropriate curves less the volume difference for the previous interval, W_t, $W_{\Delta\theta}$ and W_e can be evaluated from

$$W_t = 2\gamma|\cos\theta_i| \sum_1^n \left(\frac{V_t}{\bar{r}}\right)_n \tag{12.10}$$

$$W_{\Delta\theta} = 2\gamma(|\cos\theta_i| - |\cos\theta_e|) \sum_1^n \left(\frac{V_{\Delta\theta}}{\bar{r}}\right)_n \tag{12.11}$$

and

$$W'_e = 2\gamma|\cos\theta_i| \sum_1^n \left(\frac{V_{\Delta\theta}}{\bar{r}}\right)_n \tag{12.12}$$

The values of the work terms for numerous samples, calculated from equations (12.10), (12.11), and (12.12), show excellent agreement [108] with the corresponding work terms obtained by graphical integration of the areas in the intrusion–extrusion cycles.

12.4 THEORY OF POROSIMETRY HYSTERESIS

From the previous discussion it can be shown that a complete intrusion–extrusion cycle is composed of the following steps:

(a) Intrusion with contact angle θ_i into pores with area A_i, for which the corresponding work, W_i, is

$$W_i = \gamma |\cos\theta_i| A_i \qquad \text{(cf.10.20)}$$

(b) The change in contact angle from θ_i to θ_e as extrusion commences, for which the corresponding work, $W_{\Delta\theta}$, is

$$W_{\Delta\theta} = \gamma(|\cos\theta_i| - |\cos\theta_e|) A_e \qquad \text{(cf.12.6)}$$

By adding the area of entrapped mercury, A_t, to A_e in the above equation, the work associated with the contact angle change, $W'_{\Delta\theta}$, becomes the work necessary to alter the contact angle over the entire length of the pore, not only that portion which extrudes. Thus, $W'_{\Delta\theta}$ can be expressed as

$$W'_{\Delta\theta} = \gamma(|\cos\theta_i| - |\cos\theta_e|) A_i \qquad (12.13)$$

(c) Extrusion with contact angle θ_e is accompanied by an amount of work, W_e, given by

$$W_e = \gamma |\cos\theta_e| A_e \qquad \text{(cf.12.8)}$$

(d) After completion of extrusion and at the start of another intrusion process the entrapped mercury undergoes a contact angle change from θ_e back to θ_i, for which the corresponding work $W_{-\Delta\theta}$ is

$$W_{-\Delta\theta} = \gamma(|\cos\theta_i| - |\cos\theta_e|) A_t \qquad (12.14)$$

(e) The work, W_t, associated with mercury entrapment in the pores with contact angle θ_i is given by

$$W_t = \gamma |\cos\theta_i| A_t \qquad \text{(cf.12.5)}$$

Summing all of the above work terms (a)–(e), with positive work being done on the system and the negative sign indicating work done by the system, leads to the total work $\oint dW_1$ around a first intrusion–extrusion cycle. That is,

$$\oint dW_1 = W_i + W_{-\Delta\theta} - W'_{\Delta\theta} - W_e - W_t = 0 \qquad (12.15)$$

For second or subsequent cycles, when no entrapment occurs, W_t vanishes and the total work through the cycle becomes

$$\oint dW_2 = W_t = 0 \tag{12.16}$$

It is evident that the thermodynamic processes associated with mercury porosimetry are far more complex than just the consideration of $P-V$ work.

In volume versus radius plots for second cycles, during which no entrapment occurs, hysteresis vanishes when the correct contact angles are employed, as shown in Fig. 12.2. Hysteresis, exhibited in a first intrusion–extrusion cycle, is due solely to entrapment of mercury when the correct contact angles are employed.

The $P-V$ work differences between intrusion and extrusion, corresponding to the area between curves A and C of Fig. 12.1 can be expressed as

$$W_i - W_e = P_i V_i - P_e V_e = \gamma(|\cos\theta_i|A_i - |\cos\theta_e|)A_e \tag{12.17}$$

When the additional work terms W_t, $W'_{\Delta\theta}$ and $W_{-\Delta\theta}$ are considered the difference in the work of intrusion and extrusion is written as

$$W_i - W_e = (P_i V_i + W_{-\Delta\theta}) - (P_e V_e + W'_{\Delta\theta} + W_t) = 0 \tag{12.18}$$

Equation (12.18) establishes that the hysteresis energy, W_H, the energy associated with the area in the hysteresis region of an intrusion–extrusion cycle, is given by

$$W_H = W'_{\Delta\theta} + W_t - W_{-\Delta\theta} = \gamma(|\cos\theta_i|A_i - |\cos\theta_e|A_e) \tag{12.19}$$

Since $\oint dW = 0$ and since no work difference exists between W_i and W_e for any element of the intrusion–extrusion curve, it follows that hysteresis on volume–pressure plots results only as a consequence of ignoring the additional energy terms: $W'_{\Delta\theta}$, W_t and $W_{-\Delta\theta}$.

12.5 PORE POTENTIAL

The fact that hysteresis is attributed to the $W'_{\Delta\theta}$, W_t and $W_{-\Delta\theta}$ terms as shown in the preceding section does not, however, explain the processes which lead to these terms. Lowell and Shields [109] have postulated that, in volume versus pressure plots, hysteresis is due to the influence of a pore potential.

The concept of a pore potential is generally accepted in gas adsorption theory to account for capillary condensation at pressures well below the expected values. Gregg and Sing [110] described the intensification of the attractive forces acting on adsorbate molecules by overlapping fields from the pore wall. Adamson [111] has pointed out that evidence exists for changes induced in liquids by capillary walls over distances in the order of a micrometer. The Polanyi potential theory [112] postulates that molecules can 'fall' into the potential field at the surface of a solid, a phenomenon which would be greatly enhanced in a narrow pore.

In mercury porosimetry, it was proposed [109] that the pore potential prevents extrusion of mercury from a pore until a pressure less than the extrusion pressure is reached. Similarly, the pore potential, when applied to gas adsorption, is used to explain desorption at lower relative pressures than adsorption for a given quantity of condensed gas.

In mercury porosimetry, the pore potential can be derived as follows: The force, F, required for intrusion into a cylindrical pore is given by

$$F = 2\pi r \gamma |\cos\theta_i| = P_i \pi r^2 \tag{12.20}$$

If the pore potential, U, is the difference between the interaction of the mercury along the total length of all pores with radius r when the pores become filled at pressure P_i and when partially emptied at P_e, U can be expressed as

$$U = \int_0^{\ell_i} F d\ell_i - \int_0^{\ell_e} F d\ell_e \tag{12.21}$$

where ℓ_i represents the total length of the mercury columns in all the pores when filled and ℓ_e is the total length of the mercury which extruded from these pores.

Expressing force as pressure times area for pores in a radius interval with mean radius $\bar{r}$ and mean length $\bar{\ell}$, equation (12.21) can be rewritten as

$$U = \pi \bar{r}^2 (P_i \bar{\ell}_i - P_e \bar{\ell}_e) \tag{12.22}$$

In terms of the intruded and extruded volumes in a radius interval,

$$U = P_i V_i - P_e V_e \tag{12.23}$$

Thus, the expression for the pore potential is identical to the pressure–volume work differences between intrusion and extrusion and can be expressed as the hysteresis energy, that is,

$$-U = W'_{\Delta\theta} + W_t - W_{-\Delta\theta} = W_H \tag{cf.12.19}$$

Equation (12.19) predicts that changes in U, the pore potential, will effect the quantity of entrapped mercury and/or the difference in contact angle between intrusion and extrusion. Hence, changes in the pore potential will alter the size of the hysteresis loop.

The model described by equation (12.19) is that mercury, when intruded into a pore with contact angle θ_i, acquires an increased interfacial free energy. As the pressure decreases mercury will commence extruding from the pore at pressure P_e, reducing the interfacial area and simultaneously the contact angle, thereby spontaneously decreasing the free interfacial energy. Mercury will continue leaving the pore with the extrusion contact angle, θ_e, until the interfacial free energy equals the pore potential at which point extrusion ceases. That is, upon depressurization, mercury separates from the pore with contact angle θ_e, leaving the trapped portion of mercury near the pore opening rather than at the base of the pore. This is consistent with the observation that penetrated samples show discoloration due to finely divided mercury at the pore entrance.

Evidence for the pore potential was experimentally obtained [109] from mercury intrusion–extrusion data by impregnating various samples with both polar and non-polar materials. Fig. 12.3 is typical porosimetry data of a sample coated with various amounts of a polar material.

It is evident from Fig. 12.3 that, as the copper sulfate concentration in the sample is increased, the hysteresis increases, that is the difference between P_i and P_e increases while, at the same time, the extrusion contact angle decreases. Similarly, the work of entrapment, W_t, increases as the salt concentration is raised, as evidenced by the quantities of mercury entrapped in the sample. It can be seen in Fig. 12.3 that the intrusion curves for both treated and untreated samples are virtually identical, indicating that impregnation does not significantly alter the radius of the pore opening. However, in all cases the volume of mercury intruded decreased with increasing salt concentration, an indication that precipitation of the salt occurred near the base of the pores. These pore volume differences were eliminated in Fig. 12.3 by normalizing the data to the maximum volume intruded into the untreated sample.

The effect of pore impregnation with nonpolar material was studied by treating samples with dichlorodimethylsilane (DCDMS). In each case a decrease in hysteresis area, compared to the untreated material, was observed after coating samples with DCDMS. The increases in the extrusion contact angle, with DCDMS compared to untreated sample, resulted in decreases in $W_{\Delta\theta}$. In some cases, impregnation with DCDMS led to greater mercury retention or an increase in W_t over the untreated material. However, this was always accompanied by a larger decrease in $W_{\Delta\theta}$ and thus a decrease in the pore potential.

In some instances, it is possible to plot mercury intrusion curves corresponding to intrusion into pores within samples with which mercury amalgamates. Presumably a thin oxide film or some activation process inhibits the rate of amalgamation so that intrusion can occur as the pressure is increased

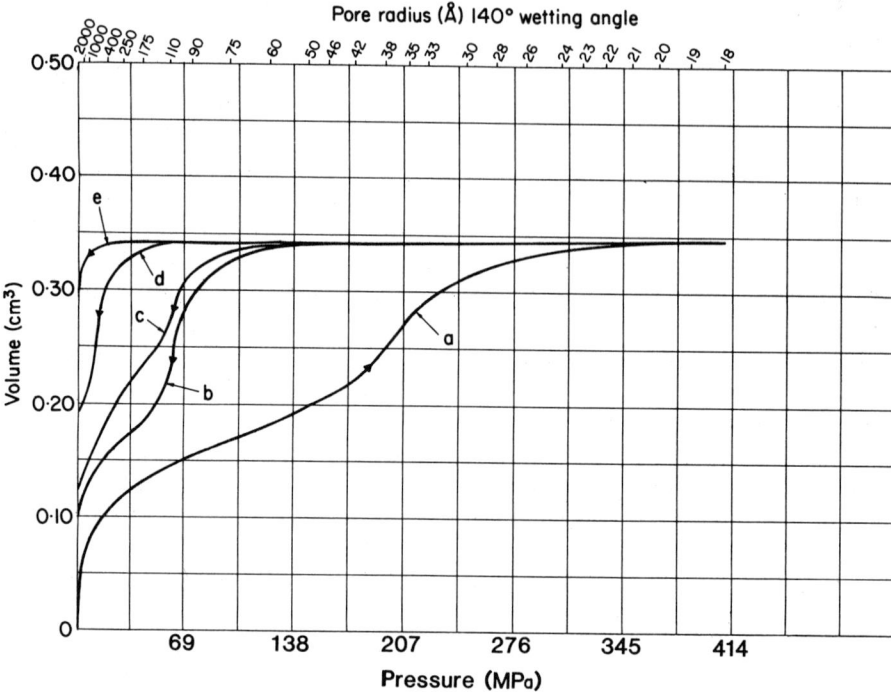

Figure 12.3 Intrusion–extrusion curves on an alumina sample coated with various amounts of copper sulfate: (a) Intrusion curve for all samples; (b) Extrusion curve for untreated alumina; (c) Extrusion curve for alumina treated with 0.5% $CuSO_4$; (d) Extrusion curve for alumina treated with 2% $CuSO_4$; (e) Extrusion curve for alumina treated with 40% $CuSO_4$.

before the unfilled pores are degraded. In these cases, the depressurization curve is always a straight horizontal line indicating that no extrusion occurs. Thus, the hysteresis energy W_H, is a maximum corresponding to the largest possible value of W_t, namely $W_t = W_i$, and corresponding to the largest possible value of $W_{\Delta\theta}$ since the extrusion angle would be that of a wetting liquid or near 0°.

12.6 OTHER HYSTERESIS THEORIES

Intrusion–extrusion hysteresis has been attributed [113] to 'ink-bottle' shaped pores. In pores of this type intrusion cannot occur until sufficient pressure is attained to force mercury into the narrow neck, whereupon the entire pore will fill. However, on depressurization the wide pore body cannot empty until a lower pressure is reached, leaving entrapped mercury in the wide inner portion. The 'ink-bottle' model ignores several factors which may reduce it to an untenable concept [114]. These include the following:

(a) All porous samples exhibit hysteresis. This would require that every porous material contains pores which are 'ink-bottle' in shape.
(b) Porosimetry curves exhibit various shapes. If hysteresis were caused by 'ink-bottle' pores only one shape of hysteresis curve should be observed.
(c) Regardless of the maximum pressure attained, depressurization always results in hysteresis. This would imply that 'ink-bottle' pores are distributed over the entire range of pore sizes. Therefore, pores with very wide entrances would have to possess even wider inner cavities.
(d) Intrusion into and extrusion out of the volume between packed spheres, where the openings are wider than the interior, show hysteresis.
(e) The 'ink-bottle' model ignores the question of the energy required to break the mercury column in the pore in order for the narrow entrance to empty while the inner cavity remains filled.

In those cases where pores are 'ink-bottle' in shape, a method, proposed by Reverberi, Feraiolo and Peloso [115] for calculating the sizes of the narrow and wide portions of the pore from intrusion–extrusion curves can be used. The method involves scanning the hysteresis loop by means of a series of pressurization and partial depressurization cycles in order to determine the volume of the wide inner portion of pores having neck radii in various radius intervals.

Androutsopoulos and Mann [116] have proposed a stochastic theory based on a two-dimensional network of interconnecting pores with varying radii to explain hysteresis and entrapment in porosimetry. By assuming that filling of some of the larger radii pores is delayed until surrounding smaller pores are filled, a mechanism similar to the filling of 'ink-bottle' pores, Androutsopoulos and Mann's calculated porosimetry curves often approximate those from actual samples.

12.7 EQUIVALENCY OF MERCURY POROSIMETRY AND GAS ADSORPTION

Lowell and Shields [117] have shown that vapor condensation–evaporation and mercury intrusion–extrusion into and out of pores are thermodynamically equivalent processes.

A vapor will condense into pores of radius r according to the Kelvin equation, viz.,

$$\ln\frac{P_v}{P_0} = \frac{-2\gamma\bar{V}\cos\theta}{rRT} \qquad \text{(cf.8.1)}$$

132 Hysteresis, entrapment and contact angle

The molar free energy change associated with the isothermal vapor pressure change from P_0 to P_v is given by

$$\Delta G = RT \ln \frac{P_v}{P_0} = \frac{-2\gamma \bar{V} \cos\theta}{r} \tag{12.24}$$

The process of mercury intrusion requires the application of hydraulic pressure, P_h, to force mercury into pores for which the molar free energy change is given by

$$\Delta G = \bar{V} \int_0^{P_h} dP_h = \bar{V} P_h \tag{12.25}$$

Combining equations (12.24) and (12.25) results in the Washburn equation, that is,

$$P_h r = -2\gamma \cos\theta \tag{12.26}$$

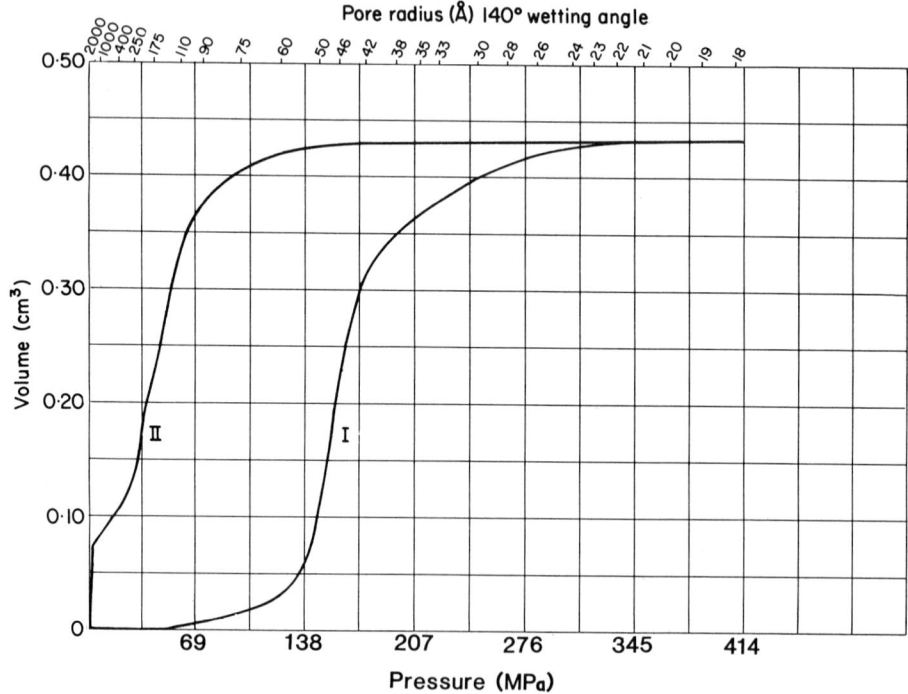

Figure 12.4 Mercury intrusion (I) and extrusion (II) curves of an alumina sample.

Equating the molar free energy terms in (12.24) and (12.25) affords an expression which relates the hydraulic pressure P_h required to force mercury into pores to the relative pressure, P_v/P_0, exerted by the liquid with radius of curvature, r. That is,

$$P_h = \frac{RT}{\bar{V}} \ln \frac{P_v}{P_0} \qquad (12.27)$$

Porosimetry isotherms, corresponding to condensation and evaporation, have been constructed [117] by conversion of the hydraulic pressure to the corresponding relative pressure using equation (12.27). A typical isotherm from mercury intrusion–extrusion data (Fig. 12.4) is shown in Fig. 12.5 for an alumina sample.

Following the convention in gas adsorption–desorption isotherms, the mercury isotherm, illustrated in Fig. 12.5, is plotted as volume versus relative pressure so that the radius increases from left to right. Curve I in Fig. 12.5 represents the condensation isotherm from the extrusion curve and curve II is the evaporation isotherm from the intrusion data. Since no adsorption takes

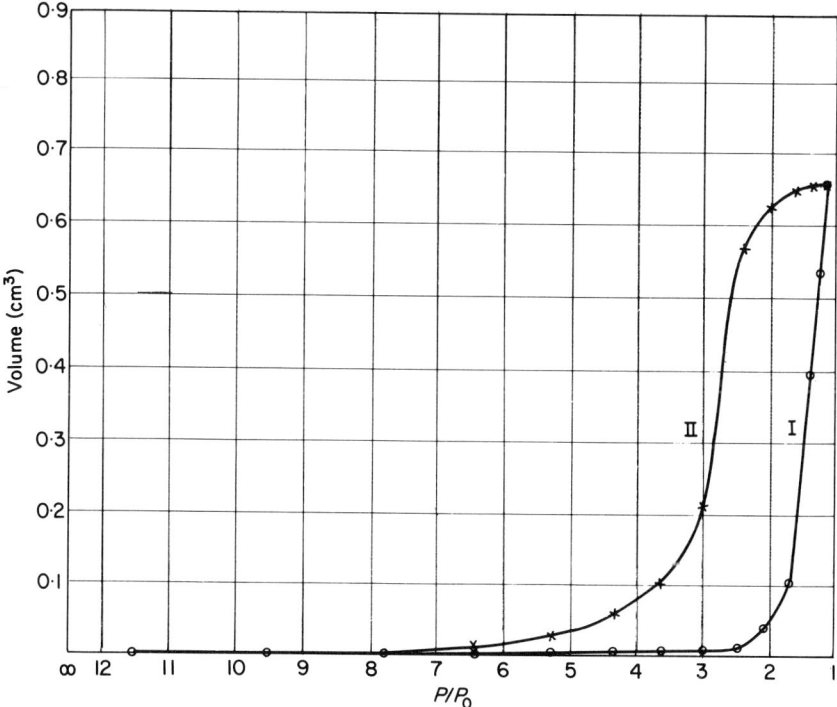

Figure 12.5 Mercury porosimetry isotherm.

place on the pore walls prior to the filling of pores in mercury porosimetry as occurs in gas adsorption, the usual 'knee' of the isotherm is absent. However, condensation–evaporation isotherms from mercury porosimetry are strikingly similar to adsorption–desorption isotherms at relative pressures above the 'knee'. The maximum volume on the intrusion curve, I in Fig. 12.4, is plotted as zero volume on the isotherm (Fig. 12.5) to conform to the requirement that the volume increases with increasing pore radius.

An advantage of isotherms constructed from mercury intrusion–extrusion curves is the capability of extending the isotherm well beyond the limits of vapor adsorption–desorption isotherms. Intruded and extruded volumes can be measured for pores of several hundred micrometers in diameter at pressures below one PSIA.

A significant aspect of the equivalency of mercury porosimetry and gas adsorption is the implication that mercury can fill pores by either liquid or vapor transport.

13
Particle size

The most common technique for measuring the particle size distribution of a fine powder is the monitoring of the change in concentration of a sedimenting suspension.

13.1 INTRODUCTION

Two methods are commonly employed for determining particle size distributions by sedimentation. The line-start technique involves the introduction of powder as a thin layer on top of a liquid medium and subsequently monitoring the particle concentration at some time and depth below the surface as the particles sediment. This method is frequently unsatisfactory due to the higher density layer of particles often settling en masse. The second method requires that the particles be uniformly suspended in a fluid medium, which is accomplished automatically by a circulating pump. The decrease in particle concentration is subsequently monitored as a function of time and depth below the surface within a sample cell.

The concentration of a homogeneous suspension of particles can be determined by their attenuation of a beam of low energy X-rays which scans the sedimentation vessel from bottom to top. The particle size is obtained from Stokes' Law as a function of particle and fluid densities, fluid viscosity and settling velocity of the particles.

13.2 STOKES' LAW

A particle sedimenting in a viscous medium under the influence of gravity is acted upon by three forces. These are the force of gravity (F_g) directing the particle down, a buoyant force (F_b) and a drag force (F_d) both acting upwards. The resultant force acting to accelerate the particle is

$$m\,dv/dt = F_g - (F_d + F_b) \tag{13.1}$$

or

$$m\,dv/dt = mg - m'g - F_d \tag{13.2}$$

where m is the particle mass, g is the gravitational acceleration and m' is the mass of fluid displaced by the particles and which therefore has a volume equal to that of the particle. The term F_d is a function of the particle velocity, v, and is zero when m is equal to m'. When a particle is less dense than the fluid, $m < m'$, the particle will rise in the liquid medium and the drag force is directed down. When $m > m'$, the particle settles in the medium and the drag force is directed upward. Stokes assumed that the drag force F_d, for spherical particles sedimenting in a viscous medium, accounts for the drag on the particle, as shown in equation (13.3):

$$F_d = 3\pi\eta Dv \qquad (13.3)$$

where η is the fluid viscosity, v is the terminal settling velocity of the particles and D is the particle diameter. Combining equations (13.2) and (13.3) yields

$$m\,dv/dt = (m - m')g - 3\pi\eta D\frac{dh}{dt} \qquad (13.4)$$

where h is the distance the particle has settled in time t.

By assuming that all particles are spherical, the mass terms m and m' in equation (13.4) can be replaced by $(\pi D^3 \rho_s)/6$ and $(\pi D^3 \rho_f)/6$ where ρ_s and ρ_f are the particle and fluid densities, respectively. Thus,

$$\left(\frac{\pi}{6}D^3\rho_s\right)\frac{dv}{dt} = \frac{\pi}{6}(\rho_s - \rho_f)D^3 g - 3\pi\eta D\frac{dh}{dt} \qquad (13.5)$$

When particles have attained their terminal velocity the term dv/dt vanishes leaving

$$\int_0^h dh = \frac{(\rho_s - \rho_f)D^2 g}{18\eta}\int_0^t dt \qquad (13.6)$$

Then

$$D = \left[\frac{18\eta}{(\rho_s - \rho_f)g}\frac{h}{t}\right]^{\frac{1}{2}} \qquad (13.7)$$

The particle size calculated by Stokes' equation (13.7) assumes spherical particle geometry. For other shapes, the diameter D is referred to as the equivalent spherical diameter, that is, the diameter of a particle having the same terminal velocity as a sphere of the same density.

13.3 X-RAY ABSORPTION

Like other forms of radiation, X-rays are absorbed in a first-order process. However, unlike visible or near-visible light, the absorption process is not unduly complicated by various scattering, refraction and diffraction processes. In fact, the absorption of visible light by fine particles is highly dependent on the wavelength of the light, the size, shape and the dielectric, as well as other properties of the particles. Rarely, if ever, is the absorptivity independent of the particle size. In the case of X-rays, however, absorption is dependent only on the mass of the absorbing material in the beam and is independent of the particle size.

The change in X-ray beam intensity with penetration distance x through an absorbing medium of density ρ can be written as:

$$\frac{-dI}{I} = (\mu/\rho)\rho dx \tag{13.8}$$

where I is the intensity of the X-ray beam and μ is the linear mass absorption coefficient which denotes the fractional beam attenuation per unit distance of penetration through the absorbing medium. The term (μ/ρ) is the mass absorption coefficient and indicates the absorption per unit mass. Integration of equation (13.8),

$$-\int_{I_0}^{I} \frac{dI}{I} = (\mu/\rho)\rho \int_0^x dx \tag{13.9}$$

gives

$$\ln I/I_0 = -\mu x \tag{13.10}$$

or

$$I = I_0 e^{-\mu x} \tag{13.11}$$

where I_0 is the initial beam intensity and I is the beam intensity after it has penetrated a distance x through the sample.

The mass absorption coefficient (μ/ρ) is proportional to the third power of the atomic number Z of the absorbing medium and the five-halves power of the X-ray wavelength (λ). That is,

$$\mu/\rho = cZ^3\lambda^{\frac{5}{2}} \tag{13.12}$$

An X-ray source with adjustable beam energy, usually between 12.5 and 15 kilovolts, provides wavelengths sufficiently large to give good absorption characteristics for all elements of atomic number 13 and higher. Thus, the need for excessive particle concentration is avoided by using longer wavelength or lower energy X-rays.

It is unnecessary to employ a monochromatic X-ray beam since each wavelength is absorbed according to equation (13.10). It follows, therefore, that if a plot of $\ln I/I_0$ versus (μ/ρ) is linear for one wavelength, it will be correspondingly linear for all wavelengths.

13.4 PARTICLE SIZE DISTRIBUTIONS

The primary measurement in an X-ray sedimentation analysis is the distribution of particle mass, as the percentage of mass finer than the indicated particle size. From this primary data, the particle number and surface area distributions are calculated.

Consider a thin element in a uniform suspension of particles at a specific depth in a sample cell through which an X-ray beam is transmitted. When sedimentation is initiated, all particles leaving the thin volume element are exactly compensated for by those entering the element of volume. Therefore, no concentration change can occur within the volume element until the largest particles located at the top of the suspension have sedimented through the volume element. When no particles of size D are any longer above the volume element to replace those that leave, the concentration within the volume element starts to decrease and becomes equal to the concentration of particles smaller than diameter D. In this case, D represents the diameter of particles falling with velocity h/t.

When sedimentation commences, that is, at time $t = 0$, the suspension will contain a uniform concentration given by:

$$C(h, 0) = \frac{(m_0)_s}{V_s + V_f} \tag{13.13}$$

where $(m_0)_s$ is the mass of suspended solid at $t = 0$, V_s and V_f are the solid and fluid volumes, respectively. At some later time t, the concentration at depth h is given by

$$C(h, t) = \frac{m_s}{V'_s + V'_f} \tag{13.14}$$

where m_s and V'_s are the mass and volume of the solid in volume V'_f of fluid measured at depth h from the top of the fluid at time t. Recognizing that the total volume of solid and fluid in the volume element is necessarily constant,

$$\frac{C(h, t)}{C(h, 0)} = \frac{m_s}{(m_0)_s} \tag{13.15}$$

Particle size distributions 139

A plot of $C(h, t)/C(h, 0) \times 100$ versus D yields a size distribution of the cumulative percentage undersize by weight.

The relationship between the size of particles and their absorption of X-rays can be derived. The absorption of X-rays by a suspension of solid in a fluid can be written as

$$\ln I/I_0 = -(\mu_f x_f + \mu_s x_s) \tag{13.16}$$

where x_f and x_s are the linear distances which the X-ray beam traverses when passing through the fluid and solid, respectively. The linear distance is proportional to the volume of solid and liquid in the volume element traversed by the beam. Thus,

$$\ln I/I_0 = -(\mu_f K V_f + \mu_s K V_s) \tag{13.17}$$

If the total volume element traversed by the beam is V, then equation (13.17) can be written as

$$\ln I/I_0 = -K[\mu_f(V - V_s) + \mu_s V_s] \tag{13.18}$$

Then

$$\ln I/I_0 = -K(\mu_f V - \mu_f V_s + \mu_s V_s) \tag{13.19}$$

Equation (13.19) establishes that when no solid is present, the X-ray absorption is due solely to the fluid, that is,

$$\ln I/I_0 = -K/\mu_f V \tag{13.20}$$

and when no fluid is present, the absorption is due only to the solid, that is,

$$\ln I/I_0 = -K\mu_s V_s \tag{13.21}$$

As the volume of solid in the beam varies during sedimentation, the term $\ln I/I_0$ varies linearly with solid volume. Rewriting equation (13.19) as

$$\ln I/I_0 = -K\mu_f V + K(\mu_f - \mu_s)V_s \tag{13.22}$$

and since for any liquid and solid μ_f and μ_s are constants, one can write

$$\ln I/I_0 = -CV + C'V_s \tag{13.23}$$

Recognizing that $V_s = m_s/\rho_s$ and that V and ρ_s are also constants yields

140 Particle size

$$\ln I/I_0 = -A + Bm_s \tag{13.24}$$

which is an equation of a straight line and therefore, establishes that as sedimentation takes place the term $\ln I/I_0$ varies linearly with solid mass remaining in the beam.

It is unnecessary to actually measure I_0. Instead, by circulating pure fluid through a sample cell the flux I_f leaving the cell can be measured. When solid is present, the flux I leaving the cell is also determined. In the absence of solid, equation (13.24) can be written as

$$\frac{I_f}{I_0} = e^{-A} \tag{13.25}$$

In the presence of solids, equation (13.24) becomes

$$I/I_0 = e^{-A_c Bm_s} \tag{13.26}$$

and thus,

$$\ln \frac{I/I_0}{I_f/I_0} = \ln \frac{I}{I_f} = Bm_s \tag{13.27}$$

Prior to commencing sedimentation, the particle suspension is circulated and the beam attenuation is measured. This provides the minimum flux value (I_{min}) due to absorption by the maximum concentration of particles. Using equations (13.24) and (13.25) gives:

$$\frac{I_{min}/I_0}{I_f/I_0} = \frac{I_{min}}{I_f} = \frac{e^{-A_c B(m_s)_0}}{e^{-A}} = e^{B(m_s)_0} \tag{13.28}$$

or

$$\ln \frac{I_{min}}{I_f} = B(m_s)_0 \tag{13.29}$$

Then, from equations (13.27), (13.29) and (13.15):

$$\frac{\ln I/I_f}{\ln I_{min}/I_f} = \frac{m_s}{(m_0)_s} = \frac{C(h, t)}{C(h, 0)} \tag{13.30}$$

A plot of $(\ln I/I_f)/(\ln I_{min}/I_f) \times 100$ versus D yields the cumulative percentage undersize by weight as shown in equation (13.30).

The validity of Stokes' equation (13.7) depends on several factors which include Reynolds number, particle porosity and density, wall effects, particle concentration, diffusion effects, terminal velocity and sample preparation and dispersion. These factors will be discussed in subsequent sections.

13.5 REYNOLDS NUMBER

Stokes' law assumes that the drag on particles sedimenting in a fluid is caused only by viscous forces in the fluid. The Reynolds number is a dimensionless quantity that is used to estimate when the fluid flow past the particle is not viscous or laminar. Above a critical Reynolds number, the fluid flow past a particle ceases to be streamlined but rather enters into a partially nonlaminar region and at very high Reynolds numbers the flow becomes turbulent. The Reynolds number (R) is given by

$$R = \frac{\rho_f v D}{\eta} \qquad (13.31)$$

where ρ_f and η are the fluid density and viscosity, respectively, v is the particle velocity and D is the equivalent spherical diameter. Combining Stokes' equation (13.7) and equation (13.31) gives:

$$D^3 = \frac{18\eta^2 R}{\rho_f(\rho_s - \rho_f)g} \qquad (13.32)$$

Experimentally, the maximum Reynolds number consistent with assuring laminar flow is 0.2. Thus,

$$D_M^3 = \frac{3.6\eta^2}{\rho_f(\rho_s - \rho_f)g} \qquad (13.33)$$

where D_M is the maximum diameter which can be analyzed with the assurance of laminar flow, that is, with a Reynolds number of 0.2.

For maximum accuracy, the largest particle diameter to be analyzed should conform to the requirement that R not exceed 0.2. However, errors of only a few (<5%) per cent are generated using R values as high 1.0. If the Reynolds number is excessively high for the desired analysis, a more viscous fluid should be chosen.

With Reynolds numbers near 0.2, irregularly-shaped particles sediment with a random orientation since the flow around the particles is laminar. At higher Reynolds numbers, the particles will assume a preferred orientation which provides the maximum drag resistance. Therefore, the particle size distribution will appear finer as the Reynolds number increases, for example, by decreasing the fluid viscosity. In the laminar flow region, the random particle orientation of irregularly-shaped particles will induce particles of any size to settle both slower and more rapidly than a sphere with the same mass. This will tend to offset the effect of irregular shape and bring the measured distribution closer to that of an equal mass of true spheres.

13.6 PARTICLE POROSITY AND DENSITY

When the particles to be analyzed contain pores, it is necessary that the sedimentation fluid wet the solid adequately to penetrate the pores. If air is entrapped in the pores, the effective particle density will be less than the true density, resulting in measured diameters smaller than the correct value. Since a dispersing agent is generally empoyed for particle dispersion this usually ensures adequate wetting of the particles so that the liquid will penetrate the pores.

For accurate particle size distributions, the correct density values should be used in Stokes' equation. Generally, densities are measured by helium pycnometry (see Chapter 22). However, for greater accuracy the measurement can be carried out by fluid displacement using the same liquid and dispersant which will be employed in the sedimentation analysis.

13.7 WALL EFFECTS

A particle settling near and parallel to a wall will be influenced by the wall proximity. The design of the sample cell should be such that particles near the side walls constituting the narrow ends of the cell are not sampled by the X-ray beam since the beam is made narrower than the windows and the side walls lie outside the window area. In the narrow dimension, however, from window to window, the sample cell is usually suffficiently narrow to influence the sedimentation rates.

The rate of sedimentation is decreased near a wall as a function of both wall proximity and particle size. The velocity v' near a wall relative to the Stokes' terminal velocity v is given by:

$$v'/v = \frac{1}{1 + \frac{9D}{32w}} \qquad (13.34)$$

where D is the equivalent spherical diameter and w is the distance of the particle from a wall. The average velocity $\bar{v}$ of a particle between the window and a plane through the center of the sample cell is given by:

$$\bar{v} = \frac{v}{w} \int_0^L \frac{dw}{(1 + 9D/32)} \qquad (13.35)$$

where L is the distance between the center plane to a wall. Upon integration, equation (13.35) becomes

$$\bar{v} = v\left[1 - \frac{9D}{32L}\ln\left(1 + \frac{32L}{9D}\right)\right] \tag{13.36}$$

For a cell 0.3 cm thick, L is 0.15 cm and equation (13.36) reduces to

$$\frac{\bar{v}}{v} = 1 - \left[1.875D\ln\left(1 + \frac{0.5333}{D}\right)^{\frac{1}{2}}\right] \tag{13.37}$$

Since, from Stokes' equation, the particle diameter is proportional to the square root of the sedimentation velocity, that is,

$$v^{\frac{1}{2}} = KD \tag{13.38}$$

equation (13.36) can be expressed as

$$\frac{\bar{D}}{D} = 1 - \left[1.875D\ln\left(1 + \frac{0.5333}{D}\right)^{\frac{1}{2}}\right] \tag{13.39}$$

where $\bar{D}/D$ is the measured diameter relative to the true Stokes' diameter.

Table 13.1 shows the error due to wall effects for various particle diameters using equation (13.39).

13.8 PARTICLE CONCENTRATION

At extremely high particle concentrations, the sedimentation rate is limited by the rate at which the liquid can flow through the powder as agglomerated particles settle en masse. With dilution, this permeation effect is no longer limiting but particle interactions and perturbations will hinder the settling rate producing incorrect size distributions.

The effect of concentration on the measured distribution can be determined by performing successive analyses on progressively more dilute suspensions. Movement of the mass distribution curve toward finer sizes indicates that the initial concentration was excessive and that the coarser particles are hindered by their passage through the slower moving finer particles. This effect is referred to as 'hindered settling'.

It must be emphasized that the particle concentration in the suspension varies with time. Usually within seconds after the start of an analysis the largest particles, and therefore a substantial percentage of the mass, will have settled out of the suspension. Thus, the particle concentration near the top of the suspension becomes more dilute with time. Therefore, concentrations

Table 13.1 Wall effects

D (micrometers)	$\bar{D}/D$	Error %	Correction (micrometers)
0.1	0.99990	0.010	0.00001
0.2	0.99981	0.019	0.00004
0.4	0.99964	0.035	0.00014
1	0.99919	0.080	0.00080
2	0.99852	0.148	0.0030
5	0.99673	0.327	0.0164
10	0.99409	0.590	0.0591
20	0.98946	1.05	0.211
50	0.97782	2.21	1.11
100	0.96182	3.82	3.82
200	0.93568	6.43	12.9

which are initially too great to give accurate results will often become sufficiently dilute to give acceptable results later in the analysis.

Almost without exception, concentrations up to 0.1–0.2 volume per cent, when well dispersed, will settle with the correct Stokes' velocity. Depending upon the size distribution, concentrations up to 3 volume per cent can be measured with negligible error.

13.9 DIFFUSION EFFECTS

When particles are very small, they can be displaced by collisions with individual molecules or groups of molecules. Essentially, the instantaneous dissymmetry in molecular collisions involving very small particles can cause the particle to undergo random motion. This phenomenon is known as Brownian motion or diffusion and can be observed by microscopy. During the process of sedimentation, Brownian motion is superimposed on the settling velocity.

After starting with a homogeneous suspension, all particles of a particular size will have passed the X-ray beam at height h at time t. However, due to Brownian motion, there will not occur a sharp boundary above which no particles of that size exist, but rather a diffuse zone will develop as Brownian motion displaces particles upward in the sedimentation vessel. Fig. 13.1(a) denotes the condition that would prevail in the absence of Brownian motion or diffusion and Fig. 13.1(b) illustrates the effect of diffusion for monodispersed particles. Fig. 13.2 is an example of the effect of Brownian motion on the measured particle size distribution. Curve a illustrates the distribution in the absence of Brownian motion; curve b would be displaced toward finer particle sizes due to diffusion upward in the sedimentation vessel.

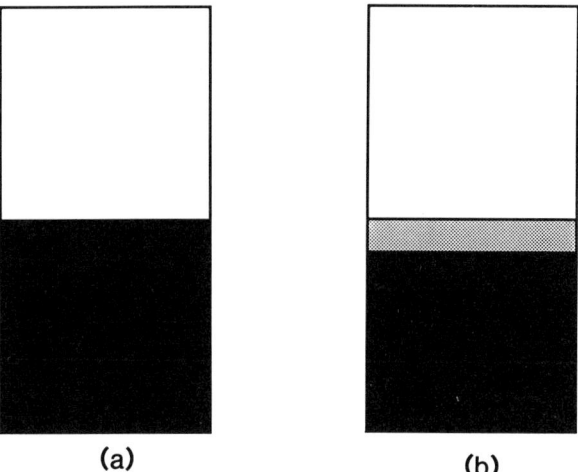

Figure 13.1 Effect of Brownian motion: (a) Absence of Brownian motion; (b) Diffusion of monodispersed particles.

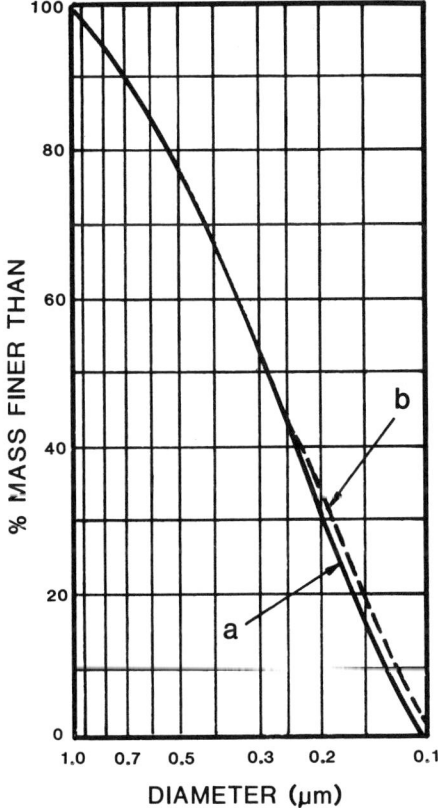

Figure 13.2 Effect of Brownian motion on particle size: (a) Particle size distribution in the absence of Brownian motion; (b) Particle size distribution in the presence of Brownian motion.

146 Particle size

The equation which describes the above diffusion process is known as Ficks' Second Law and is given by

$$\left(\frac{\delta C}{\delta t}\right)_x = \bar{D}\frac{\delta^2 C}{(\delta x^2)_t} \qquad (13.40)$$

where C is the particle concentration, t and x are the time and distance which the particles have diffused, respectively. The term $\bar{D}$ is the diffusion coefficient.

A solution which satisfies Ficks' Law for diffusion is given by

$$\frac{C}{C_0} = \frac{1}{2} - \int_0^Z P(Z)dZ = \frac{1}{2} - \frac{1}{2}\left[\frac{2}{m}\int_0^Z \exp(-Z^2)dZ\right] \qquad (13.41)$$

where

$$Z = \frac{x}{(4\bar{D}t)^{\frac{1}{2}}} \qquad (13.42)$$

and C_0 is the initial concentration in the homogeneous suspension of particles of a given size, C is the concentration at the center of the interface at time t and x is the distance from the interface.

In order to evaluate the diffusion coefficient $\bar{D}$, one must consider Ficks' first law of diffusion

$$J = -\bar{D}\frac{\delta C}{\delta x} \qquad (13.43)$$

where J is the material or particle flux across a unit area per unit of time. The term $\bar{D}$ is the diffusion coefficient or the quantity of particles migrating across a unit area per unit time per unit concentration gradient.

The driving force which causes diffusion is Gibb's partial molal free energy of the size component of interest, in this case the chemical potential of a single size particle. This force is given by

$$F_{\text{diff}} = \frac{\delta U_p}{\delta x} \qquad (13.44)$$

where U_p is the Gibb's partial molal free energy or chemical potential of the particle and x is the distance over which the driving force is applied. For any single particle one can write

$$F_{\text{diff}} = \frac{1}{N}\frac{\delta U_p}{\delta x} \qquad (13.45)$$

where N is Avogadro's number, since U is a molar quantity. For dilute solutions, the chemical potential (μ_p) can be expressed as

$$\mu_p = \mu_p^o - RT \ln C \tag{13.46}$$

where μ_p^o is the chemical potential in a reference or standard state and C is the particle concentration. The terms R and T are the molar gas constant and absolute temperature, respectively. Then,

$$\frac{\delta U_p}{\delta x} = \frac{RT}{N}\frac{d \ln C}{dx} \tag{13.47}$$

so that

$$F_{\text{diff}} = \frac{RT}{N}\frac{D \ln C}{dx} \tag{13.48}$$

Under stationary state conditions, that is, when the particle is equal in each direction, the diffusional force (F_{diff}) is equal to the force of viscous resistance given by equation (13.3)

$$F_d = 3\pi\eta Dv \tag{cf.13.3}$$

Therefore, the diffusion velocity is given by

$$v_{\text{diff}} = \frac{RT}{3\pi\eta ND}\frac{d \ln C}{dx} \tag{13.49}$$

where the term $RT/3\pi\eta ND$ is the diffusion coefficient $\bar{D}$. Substituting this value into equation (13.42) gives

$$Z = x\left(\frac{3\pi\eta ND}{4RTt}\right)^{\frac{1}{2}} \tag{13.50}$$

Using the following values:

$R = 8.314 \times 10^7$ ergs deg^{-1} mol^{-1}
$T = 35\,°C$ (308K)
$N = 6.023 \times 10^{23}$ particles mole^{-1}
$\eta = 0.007225$ poise (water 35 °C)

equation (13.50) becomes

$$Z = x(6.33 \times 10^5)(D/t)^{\frac{1}{2}} \tag{13.51}$$

It is convenient to express x, the distance from the interface at which the particle concentration is measured, as the difference between the distance, h, at which the interface would have sedimented in the absence of diffusion as a sharp boundary layer and some arbitrary distance h'. It should be recalled that

148 Particle size

diffusion from the interface is independent of sedimentation. Equation 13.51 then becomes

$$Z = (h - h')(6.33 \times 10^5)(D/t)^{\frac{1}{2}} \qquad (13.52)$$

Consider particles distributed in the range from 100 micrometers to 0.1 micrometer. Assume that they are sedimenting in water at 35 °C and that their density is 4.0 g/cm³. Columns 2 and 3 in Table 13.2 show the time and distance fallen in order for particles of various diameters to intersect the center of the X-ray beam, based on a typical analysis. Column 4 is $(D/t)^{1/2}$ and column 5 is Z calculated from equation (13.52), if the arbitrary distance h' is taken to be $(h - 0.01)$ cm. The value of C/C_0 at this distance can then be calculated from equation (13.41). Column 6 in Table 13.2 shows the values of C/C_0 using the value of h from column 3 at a depth of $(h - 0.01)$, as calculated from equation (13.41), the probability integral (erf$x$).

The nature of the probability integral is such that when $Z = 0$ at $x = 0$, the value of C/C_0 is 0.5. This results from the fact that the center of the diffuse zone will always contain 50% of the particle concentration. One may consider that even at $t = 0$ with a sharp boundary layer, each particle in the layer is 50% below and 50% above the boundary. The values of C/C_0 corresponding to those in Table 13.2 represent the concentration extending out 0.01 cm from the center of the diffuse zone.

It is evident that under the nominal conditions specified, Brownian motion

Table 13.2

Particle diameter (micrometers)	t (s)	h (cm)	D/t	Z	C/C_0
100.0	1.4017	3.1750	8.44×10^{-2}	534	0.00
50.12	5.5082	3.1750	3.00×10^{-2}	190	0.00
25.12	22.215	3.1750	1.06×10^{-2}	67.1	0.00
10.00	140.17	3.1750	2.67×10^{-3}	16.9	0.00
5.012	558.04	3.1750	9.48×10^{-4}	6.00	0.00
1.995	5249.8	1.1892	1.95×10^{-4}	1.23	0.04
0.8913	5271.1	0.9058	1.30×10^{-4}	0.823	0.12
0.7943	5285.3	0.7554	1.23×10^{-4}	0.778	0.14
0.7080	5301.9	0.6019	1.16×10^{-4}	0.734	0.15
0.6026	5327.7	0.4382	1.06×10^{-4}	0.671	0.17
0.5012	5360.6	0.3050	9.67×10^{-5}	0.612	0.19
0.3981	5395.7	0.1937	8.59×10^{-5}	0.544	0.22
0.3020	5446.1	0.1125	7.45×10^{-5}	0.472	0.25
0.1995	5506.5	0.0497	6.02×10^{-5}	0.381	0.30
0.1000	5606.7	0.0127	4.22×10^{-5}	0.267	0.35

or diffusion plays an insignificant role until particle diameters smaller than 2 micrometers are examined. The influence of Brownian motion will decrease with more dense materials because sedimentation occurs more rapidly and with increased viscosity because the diffusion rate is diminished.

In the absence of diffusion, a sharp boundary of single sized particles, when in the center of the X-ray beam, will attenuate the beam by 50% since half the beam is absorbed by 100% of the particle concentration and the other half is not absorbed by any particles. In the case of diffusion, the same 50% of the beam will be absorbed when the center of the diffuse zone is in the center of the beam. Therefore, the process of diffusion must be considered in the light of the effect which the diffusing particles have on the X-ray beam absorption when particles of one size are being measured.

For example, consider the influence of 0.3 micrometer particles when 0.2 micrometer particles are being measured. When the middle of the diffuse boundary of the 0.2 micrometer particles is centered in the X-ray beam, the center of the diffuse boundary of the 0.3 micrometer particles will be below the beam. The exact distance is given by Stokes' Law, for the conditions mentioned previously as

$$h_{0.3} - h_{0.2} = \left(\frac{(3 \times 10^{-5})^2(4.0 - 0.99406)(5506.5)(980)}{(18)(0.007225)}\right)^{\frac{1}{2}}$$
$$- 0.0497 \qquad (13.53)$$

or

$$h_{0.3} - h_{0.2} = 0.11226 - 0.0497 = 0.0626 \qquad (13.54)$$

For the purpose of the above calculation, the diameter data in Table 13.2 have been rounded off. Using the value of 0.0626 cm in equation (13.52) gives

$$Z = 0.0626(6.33 \times 10^5)\left(\frac{3 \times 10^{-5}}{5506.5}\right)^{\frac{1}{2}} = 2.92 \qquad (13.55)$$

Using this value of Z in the probability integral, equation (13.41) gives

$$C/C_0 = 0.00002$$

Thus, 0.002% of the 0.3 micrometer particles will be measured as excess mass when the 0.2 micrometer particles are measured. This percentage increases as the difference between the diameters being measured and the influencing particles become smaller. However, as the error in mass increases the error in diameter decreases.

For particles over 2 micrometers, the above analysis discloses that diffusion

150 Particle size

produces an entirely negligible error. For wide size distributions, the mass error introduced by diffusion of the very small particles also becomes negligible since this small error is introduced into the measurement of a small percentage of the total mass.

13.10 TERMINAL VELOCITY

Stoke's law is valid only if the sedimenting particles have attained their terminal velocity before their concentration is measured. It is possible to calculate the time required for an accelerating particle to reach 99.9% of its final velocity.

Restatinig equation (13.4), after replacing masses m and m' by $\rho_s V_s$ and $\rho_f V_f$, respectively, gives

$$\rho_s V_s \frac{dv}{dt} = (\rho_s V_s - \rho_f V_f)g - 3\pi\eta D \frac{dh}{dt} \tag{13.56}$$

Assuming that the particles are spherical and initially at rest, results in

$$\frac{\pi}{6}(\rho_s - \rho_f)gD^3 - 3\pi D\eta v = \frac{\pi}{6}\rho_s D^3 \frac{dv}{dt} \tag{13.57}$$

Then

$$\frac{dv}{dt} = \frac{(\rho_s - \rho_f)g}{\rho_s} - \frac{18\pi v}{\rho_s D^2} \tag{13.58}$$

Assuming that

$$\frac{\rho_s - \rho_f}{\rho_s} = A \tag{13.59}$$

and

$$\frac{18\eta}{\rho_s D^2} = B \tag{13.60}$$

then

$$\int_0^v \frac{dv}{Ag - Bv} = \int_0^t dt \tag{13.61}$$

and

$$v = \frac{Ag}{B}[1 - \exp(-Bt)] \tag{13.62}$$

or

$$v = \frac{(\rho_s - \rho_f)gD^2}{18\eta}\left[1 - \exp\left(\frac{-18\eta t}{\rho_s D^2}\right)\right] \tag{13.63}$$

Since

$$\frac{(\rho_s - \rho_f)gD^2}{18\eta} = v_t \tag{13.64}$$

where v_t is the Stokes' terminal velocity, equation (13.63) becomes

$$v = v_t\left[1 - \exp\left(\frac{-18\eta t}{\rho_s D^2}\right)\right] \tag{13.65}$$

When $v = 0.999 v_t$ one obtains

$$1 - \exp\left(\frac{-18\eta t}{\rho_s D^2}\right) = 0.999 \tag{13.66}$$

or

$$\exp\left(\frac{-18\eta t}{\rho_s D^2}\right) = 0.001 \tag{13.67}$$

Solving equation (13.67) for t gives the time required to achieve 99.9% of the terminal velocity. Table 13.2 illustrates the time for particles of various densities and diameters sedimenting in water for which η is assumed to be 0.01 poise.

It is evident from Table 13.3 that the time for a particle undergoing acceleration to reach 99.9% of its terminal velocity is in every case a small part of one second. For example, a one micrometer particle with a density of $2\,g/cm^3$ will reach this velocity in $7.68 \times 10^{-7}\,s$. This introduces a negligible error when compared to the total sedimentation time.

The distance a particle moves while accelerating to 99.9% of its terminal velocity can be calculated by rewriting equation (13.65) as follows:

$$\int_0^h dh = v_t \int_0^t \left[1 - \exp\left(\frac{-18\eta t}{\rho_s D^2}\right)\right] dt \tag{13.68}$$

which results in:

Table 13.3 Time(s) to achieve 99.9% of terminal velocity

s (g/cm^3)	0.1 μm	1 μm	10 μm	100 μm
2	7.68 × 10^{-9}	7.68 × 10^{-7}	7.68 × 10^{-5}	7.68 × 10^{-3}
3	11.5 × 10^{-9}	11.5 × 10^{-7}	11.5 × 10^{-5}	11.5 × 10^{-3}
5	19.2 × 10^{-9}	19.2 × 10^{-7}	19.2 × 10^{-5}	19.2 × 10^{-3}
10	38.4 × 10^{-9}	38.4 × 10^{-7}	38.4 × 10^{-5}	38.4 × 10^{-3}

$$h = v_t t \left(1 - \frac{\rho_s D^2}{18\eta t}\right)\left[1 - \exp\left(\frac{-18\eta t}{\rho_s D^2}\right)\right] \tag{13.69}$$

Recalling equation (13.67),

$$\exp\left(\frac{-18\eta t}{\rho_s D^2}\right) = 0.001 \tag{cf.13.67}$$

then

$$h = v_t t \left[1 - \left(\frac{\rho_s D^2}{18\eta t}\right)(0.999)\right] \tag{13.70}$$

Solving equation (13.67) for $\rho_s D^2 / 18\eta t$ and substituting into equation (13.70) gives:

$$h = 0.855 \, v_t t$$

For all cases, the distance a particle moves before achieving 99.9% of its terminal velocity is a negligibly small fraction of its sedimentation distance during analysis.

Part 2
EXPERIMENTAL

14
Adsorption measurements — preliminaries

14.1 REFERENCE STANDARDS

All equipment designed to measure surface area, adsorption–desorption isotherms or pore volume by adsorption actually determines the quantity of gas condensed on a solid surface at some equilibrium vapor pressure. The surface area or pore volumes and pore sizes are then calculated by means of an appropriate theory used to treat the adsorption and/or desorption data. Depending on the apparatus employed, the adsorbed quantity is measured as volume or weight. The accuracy of an adsorption apparatus is, therefore, dependent upon its ability correctly to measure either of these quantities.

If agreement is not reached between a properly calibrated apparatus and a reference standard, the reference material must be questioned. All too often materials used as standards for surface area or pore volume prove to be inadequate for a wide variety of reasons, some of which are listed below:

- Irreversible chemical reactions with atmospheric constituents, particularly water vapor and oxygen.
- Breaking and abrasion of particles incurred during handling.
- Diffusion to and from the surface, to and from the interior.
- Cold flow of some materials below but near their melting points.
- Particle segregation leading to nonrepresentation.
- Variations in sample conditioning.

These and other factors must be examined before a method or apparatus is condemned for lack of agreement with some other method or instrument. It is often simpler and more definitive to establish that an apparatus is properly working and calibrated than it is to use reference standards which are subject to the hazards listed above. Furthermore, the question must always arise regarding the merits of the technique and apparatus used to establish the value of the standard in the first instance. Experimental results obtained with a carefully calibrated apparatus are preferred to results which may agree with a presumed standard whose properties may change or which may have been incorrectly measured in another laboratory.

14.2 OTHER PRELIMINARY PRECAUTIONS

Other factors over which caution must be exercised when comparing measurements include gas purity and the temperature at which the data is obtained. In some instances, small quantities of impurity in the adsorbate vapor can be preferentially adsorbed. On nonporous surfaces, this will not severely alter the results. However, on porous samples, impurities can block the entrance to smaller pores and, in extreme cases, even condense in the pores, thereby drastically altering the measurement. Even when using high-purity gases a cold trap is recommended. Gases should be transferred from the pressurized tank to the apparatus with metal tubing (copper or stainless steel). Plastic tubing is not at all satisfactory because atmospheric components can often diffuse through the walls into the system. Also, many plastics contain plasticizers that can diffuse into the adsorbate as it is transferred. When metal tubing is used, it is necessary to heat the tubing to about 150°C and purge with dry nitrogen for twenty-four hours to remove lubricating residue used during the extruding process.

The temperature of the adsorbent and vapor with which it is in equilibrium must remain constant during the measurement. With nitrogen as the adsorbate, the coolant most often used is liquid nitrogen. The temperature of liquid nitrogen is dependent on ambient pressure and the presence of impurities in the liquid which tend to elevate the boiling point. The Clapeyron–Clausius [118] equation can be used to calculate the change in liquid nitrogen temperature near its normal boiling point. Using 77.36 K ($-195.79\,°C$) as the normal boiling point and 1335 cal mol^{-1} as the heat of vaporization of nitrogen gives

$$\ln P = 15.3182 - \frac{671.867}{T} \qquad (14.1)$$

Thus, near the normal boiling point the temperature will change by 0.01 K when the equilibrium pressure changes by 0.86 mm of Hg.

14.3 REPRESENTATIVE SAMPLES

Often, for the purposes of laboratory analysis, it is necessary to obtain a small quantity of powder from a larger batch. For maximum accuracy and reproducibility, it is necessary that the sample chosen be representative of the larger initial quantity. (Here, the term representative means that the sample must possess the same particle and pore size distributions and specific surface area as the larger quantity from which it was obtained.)

To some extent, under even slight agitation, particles tend to segregate with the finer ones settling toward the bottom of the container. When poured from a container into a conical pile, the smaller particles will tend toward the center and toward the base. This behavior is caused by the small particles settling through the voids between the larger ones. In essence, the larger particles with a lower bulk density can be thought of as floating on the smaller ones with a higher bulk density. (Bulk density is defined as the mass divided by the sum of the particle and void volumes.)

It is generally impossible to make a segregated sample homogeneous by shaking, tumbling or any other technique. Often these attempts only further enhance the segregation process. Devices such as the spinning rifflers shown in Figs. 14.1 and 14.2 can be used to obtain representative samples. Rifflers operate on the principle that a sample need not be homogeneous in order to be representative.

The rifflers shown in Figs. 14.1 and 14.2 operate by loading the powder sample into a vibrating hopper which delivers the sample down a chute into eight rotating collectors. Both the delivery and rotational rates can be controlled.

The sample, when loaded in the hopper, will be segregated. Therefore, at any depth ℓ there will exist a particle diameter gradient $\Delta D/\Delta \ell$. The powder settles as it is delivered to the collectors at the rate $\Delta \ell/\Delta t$. Then,

$$\frac{\Delta D}{\Delta \ell} \cdot \frac{\Delta \ell}{\Delta t} = \frac{\Delta D}{\Delta t} \qquad (14.2)$$

which is the time rate of change of particle diameters leaving the hopper. If ψ is defined as the change in particle diameter entering each collector per revolution, then,

$$\psi = \frac{1}{8} \frac{\Delta D}{\Delta t} \tau \qquad (14.3)$$

where τ is the time per revolution. Substituting equation (14.2) into (14.3) gives

$$\psi = \frac{1}{8} \frac{\Delta D}{\Delta \ell} \cdot \frac{\Delta \ell}{\Delta t} \tau \qquad (14.4)$$

Recognizing that $\Delta \ell/\Delta t$ is proportional to the feed rate F yields

$$\psi = K \frac{\Delta D}{\Delta \ell} F \tau \qquad (14.5)$$

Equation (14.5) asserts that the change in particle diameter entering each collector can be made as small as necessary by decreasing the feed rate or increasing the collector's rotational rate. However, Hatton [119] argues that it

Figure 14.1 Sieving Riffler. Courtesy of Quantachrome Corporation.

is preferable to slow the feed rate rather than increase the rotational rate to provide better representation.

When the entire sample has been delivered, each collector will contain powder exactly representative of the initial batch. Each collector will also contain a size gradient from top to bottom. Therefore, if the quantity required for analysis is less than the amount in any single collector, the process must be repeated. This is achieved by placing the contents of one collector back into the hopper and reriffling. The Micro Riffler, Fig. 14.2, can

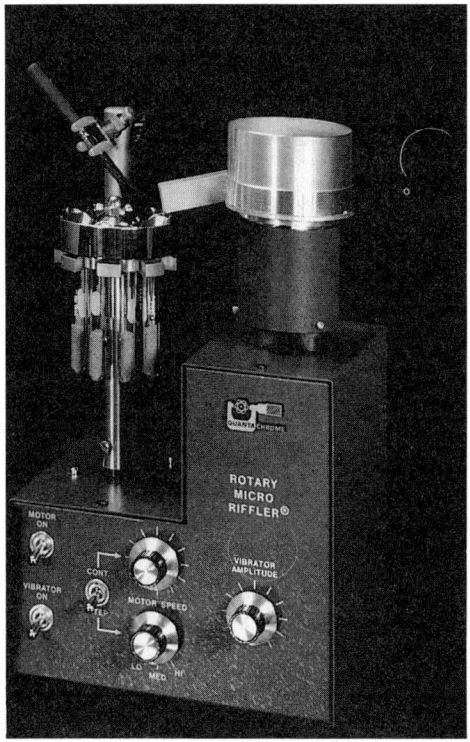

Figure 14.2 Micro Riffler. Courtesy of Quantachrome Corporation.

riffle samples of less than one gram to accommodate the need for small final samples.

The large riffler, Fig. 14.1, has an optional sieve which can be placed on top of the hopper to exclude particles above a required size. The advantage of this arrangement lies in the fact that a single particle of 100 μm radius has the same weight as one million one-micrometer particles with only one-hundredth the surface area. If only a few of the large particles are present they may not be properly represented in the final sample.

Often it is thought that the effectiveness of a riffler can be demonstrated by the uniformity of weight accumulated in each collector. This reasoning is incorrect if one considers that each collector will necessarily acquire a slightly different amount of sample if the collector diameters vary slightly. The only correct test for the effective performance of a riffler is to compare the contents of each collector in terms of particle size distribution or specific surface area.

Experiments with silica powder chosen from five depths in a two pound container gave surface areas from top to bottom of 9.8, 10.2, 10.4, 10.5, and

$10.7 \, \text{m}^2 \, \text{g}^{-1}$. When the same sample was poured into a conical pile, five random samples produced surface areas of 10.3, 11.0, 10.4, 10.0, and $10.6 \, \text{m}^2 \, \text{g}^{-1}$. However, when the sample was riffled in a spinning riffler with three size reductions, the subsequent analysis of the contents of five collectors gave 10.2, 10.1, 10.2, 10.2, and $10.1 \, \text{m}^2 \, \text{g}^{-1}$ as the specific surface area.

14.4 SAMPLE CONDITIONING

Before a sample can be analyzed it is necessary to remove contaminating materials which can alter the surface potential and block or fill pores. Sample conditioning can be accomplished by vacuum pumping or purging with an inert gas. Both methods require the use of elevated temperatures to hasten the rate at which contaminants leave the surface. Caution must be used not to reach temperatures at which the surface properties of the sample can be altered. Melting, dehydration, sintering, and decomposition are processes that can drastically alter surface properties. A test to establish that excessive degassing temperatures have not been used is to determine the surface area of a duplicate sample, after degassing at higher temperature, for a longer time. Duplication of results indicates the initial degassing conditions were satisfactory unless both analyses were performed on a completely degraded sample.

When a vacuum is used to degas, precautions must be taken to prevent the initial surge from blowing powder throughout the system. In a well-sealed system, degassing can be taken as completed when no pressure increase can be detected after isolation from the pump. For routine work generally 10^{-3}–10^{-5} torr (mm of Hg) is sufficient. The usual precautions, such as a baffle and a cold trap, should be taken to insure no contamination by the pump oil.

When degassing samples by purging, the purge gas must be of high purity and passed through a cold trap to remove traces of organic compounds or water vapor. Completion of degassing can be determined by passing the effluent over a thermal conductivity detector which can detect impurities in the effluent as low as a few parts per million. Fig. 14.3 illustrates a convenient flow system for degassing using a purge. Detector D_2 compares the effluent from the sample cell f with clean gas entering detector D_1. A reference or zero impurity signal is obtained by switching valve e to bypass the sample cell. Helium or hydrogen will serve best as the purge gas (see Section 16.3).

In those instances where samples cannot be heated, the method of repetitive cycling investigated by Lopez-Gonzales, Carpenter and Deitz [120] can be utilized. They found that by repetitive adsorption and desorption the surface can be adequately cleaned to allow reproducible measurements. Usually three to six cycles are sufficient to produce a decontaminated surface. Presumably the process of desorption, as the sample temperature is raised, results in momentum exchange between the highly dense adsorbate leaving the surface and the contaminants. As the impurities are removed from the

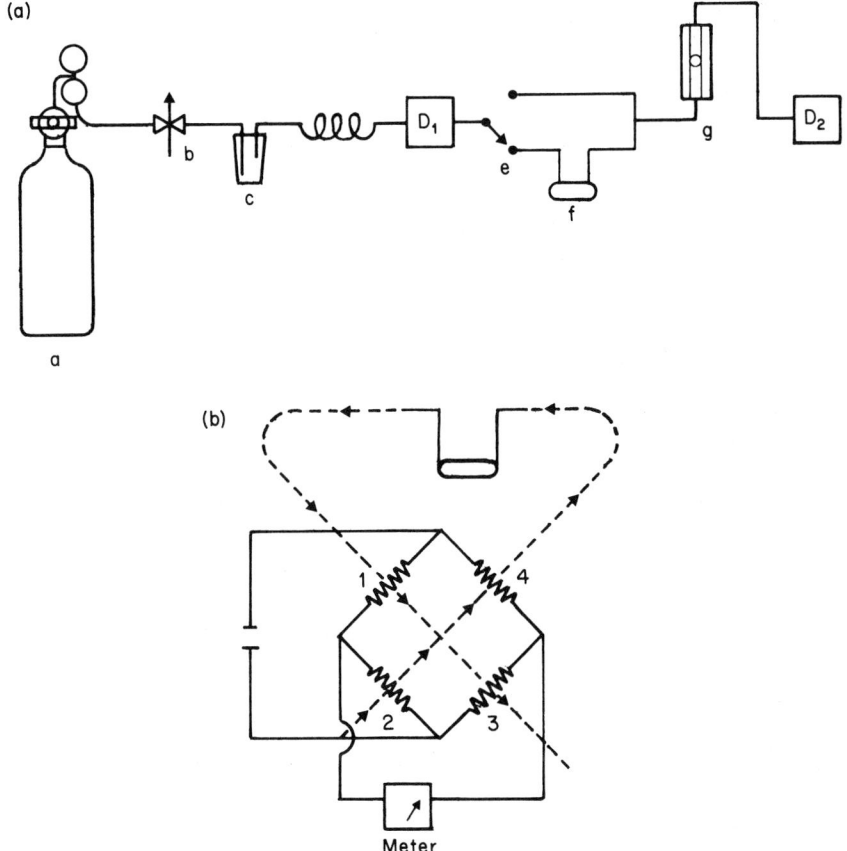

Figure 14.3 Flow outgassing with thermal conductivity detector: (a) purge gas source; b, needle valve; c, cold trap; D_1 and D_2, thermal conductivity detectors; e, flow selector valve; f, cell with powder; g, flow meter; (b) Heated filaments 1 and 3 are D_2, 2 and 4 are D_1. Dotted line is gas flow through sample cell. Signal meter is shown at bottom.

surface they will be carried out of the sample cell by the flowing gas. Thus, the technique of repetitive cycling is an efficient means for removal of contaminants from the surface of a solid. The flow diagram of Fig. 14.3 is well suited to this type of procedure.

15
Vacuum volumetric measurements

15.1 NITROGEN ADSORPTION

Many types of vacuum adsorption apparatus have been developed [121–124] and no doubt every laboratory where serious adsorption measurements are made has equipment with certain unique features. The number of variations are limited only by the need and ingenuity of the users. However, all vacuum adsorption systems have certain essential features, including a vacuum pump, two gas supplies, a sample container, a calibrated volume, manometer and a coolant.

Fig. 15.1 shows a simplified vacuum adsorption apparatus which is suitable for nitrogen adsorption.

The volumes V_a, V_b, and V_c, between fiducial marks, are carefully calibrated prior to sealing in place. Calibration is achieved by filling with mercury, emptying, and weighing. The manifold volume V_m, indicated by the striped area in Fig. 15.1, is determined by admitting helium through stopcock 2, into the evacuated manifold. Before and after admitting helium, the mercury in the left arm of the manometer is brought to the fiducial mark by adjusting stopcock 8. With helium in the manifold, the pressure P is read as the difference in mercury levels in the manometer arms. When stopcock 5 is turned, mercury is allowed to fill the lower bulb V_a by rising from the lowest fiducial mark to the one above. The manometer is again brought to its fiducial mark and pressure P' is noted. From the ideal gas equation, one obtains

$$P(V_m + V_a + V_b + V_c) = P'(V_m + V_b + V_c) \tag{15.1}$$

The manifold volume, V_m, can then be expressed as

$$V_m = \frac{PV_a - (P' - P)(V_b + V_c)}{(P' - P)} \tag{15.2}$$

The sample cell is not completely filled by the adsorbent; it contains a void volume V_v which is the volume not occupied by adsorbent, up to stopcock 4. If the calibrated volumes and manifold are filled with nitrogen at pressure P_1

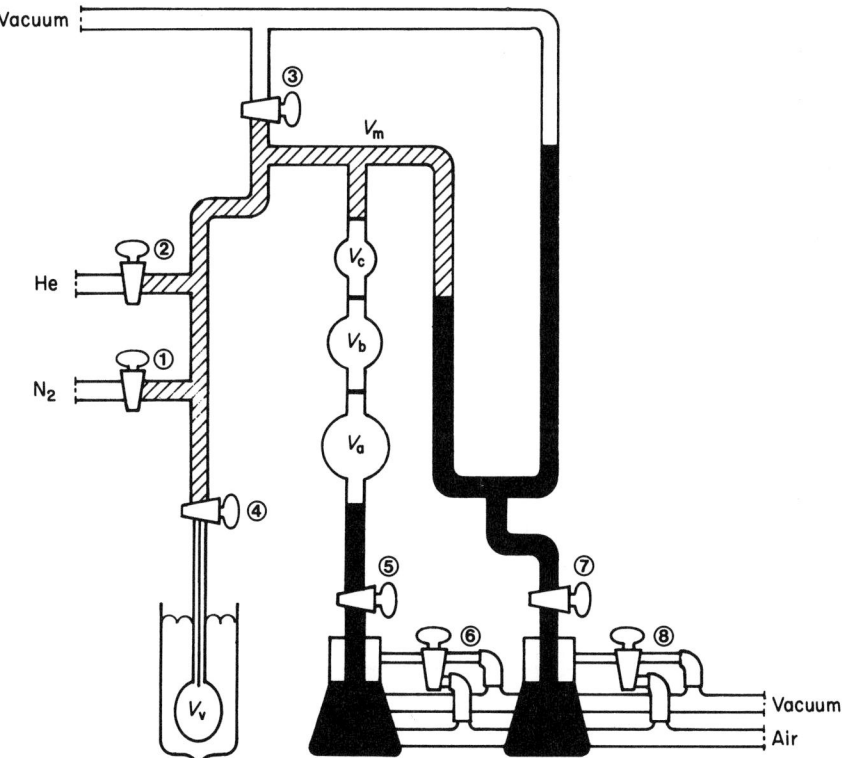

Figure 15.1 Simplified vacuum adsorption apparatus. Shaded areas represent mercury. Heavy horizontal lines are fiducial marks.

and stopcock 4 is opened the pressure will fall because of expansion and adsorption of nitrogen in the evacuated cell. After equilibrium is attained and the mercury in the manometer is returned to the fiducial mark the new pressure P_1' is noted. The number of moles n_1 adsorbed is given by

$$n_1 = \frac{P_1(V_a + V_b + V_c + V_m)}{RT} - \frac{P_1'(V_a + V_b + V_c + V_m)}{RT} - \frac{P_1'V_v}{RT_v} \tag{15.3}$$

where T is the temperature of the calibrated volumes and manifold and T_v is the temperature within the sample cell, assumed to be uniform. The value of V_v can be determined by using helium gas and repeating the preceding steps, the so-called blank run, during which helium is not significantly adsorbed at liquid nitrogen temperature. Since no adsorption occurs, n_1 in equation (15.3) is zero and V_v can be calculated.

With pressure in atmospheres and all volumes in cubic centimeters, equation (15.3) can be rewritten in terms of the volume adsorbed V_1 at standard temperature (273.2 K) and pressure (1.0 atm.)

$$V_1 = P_1 V_0^\dagger - P_1'(V_0^\dagger + V_v^\dagger) \tag{15.4}$$

Using V_0 as the sum of the calibration volumes V_a, V_b, V_c, and the manifold volume V_m and reducing all gas volumes to standard conditions yields the following:

$$V_0^\dagger = \frac{V_0}{T}\left(\frac{273.2}{760}\right) \tag{15.5}$$

$$V_v^\dagger = \frac{V_v}{T_v}\left(\frac{273.2}{760}\right) \tag{15.6}$$

For subsequent data points stopcock 4 is closed and additional nitrogen is admitted into the calibrated volumes and manifold. When stopcock 4 is opened the pressure will fall from P_2 to P_2' and the volume adsorbed under standard conditions is given by

$$V_2 = [P_1 V_0^\dagger + (P_2 - P_1')V_0^\dagger] - P_2'(V_0^\dagger + V_v^\dagger) \tag{15.7}$$

In general, for the nth step the volume adsorbed is written as

$$V_n = [P_1 V_0^\dagger + (P_2 - P_1')V_0^\dagger + \ldots + (P_n - P_{n-1}')V_0^\dagger] \\ - P_n'(V_0^\dagger + V_v^\dagger) \tag{15.8}$$

The terms in the square bracket represent the sum of the gas added to the system, and the second term contains the equilibrium quantity of gas remaining unadsorbed.

The purpose of several calibrated volumes V_a, V_b, and V_c, is to enable more than one data point to be obtained from each filling. This is accomplished by raising the mercury level to arrive at a new initial pressure before opening stopcock 4. Also, the ability to adjust the calibrated volume gives some control over the pressure difference when small or large amounts of gas are adsorbed.

Points on the desorption isotherm are obtained by first adsorbing at some relative pressure close to unity. Then, stopcock 4 is closed and the calibrated volumes and manifold are evacuated. Stopcock 4 is opened which permits desorption to occur. The equilibrium pressure is obtained from the manometer and the volumes desorbed are calculated using equation (15.8) except that P' is now the pressure after desorption.

Attempts to measure points on the desorption isotherm without first adsorbing near saturation will result in scanning the hysteresis loop from the adsorption to the desorption curve (see Fig. 16.13 and Chapter 16, Section 6).

The errors in the vacuum volumetric method arise principally from two sources. The error in measured doses is cumulative, requiring very accurate calibration. Also, the void volume correction becomes more significant at higher pressures when measuring low surface areas because more molecules are retained in the void volume than are adsorbed.

15.2 DEVIATIONS FROM IDEALITY

The gas contained in the void volume of the sample cell must be corrected for deviations from ideality. Prior to reducing the void volume V_v to standard conditions using equation (15.6), the measured volume should be corrected for nonideality. For nitrogen, Emmett and Brunauer [125] derived the appropriate correction, which is linear with pressure, from the van der Waals equation. Using V'_v as the corrected void volume, the correction for non-ideality is

$$V'_v = V_v(1 + \alpha P) \qquad (15.9)$$

where $\alpha = 6.6 \times 10^{-5}$ for nitrogen when P is expressed in torr. Table 15.1 gives α values for various gases.

15.3 SAMPLE CELLS

Sample cells should be designed to minimize the void volume. Samples with adequately low surface area may adsorb less volume than is required to fill the void volume. In such instances large errors can be generated unless the void volume is accurately measured and reduced to as small a value as possible.

Ideally, the sample should be placed in an open-ended glass cell and then closed with a torch to seal it into the adsorption apparatus.

The stem of the sample cell should be made of capillary tubing to further reduce the void volume and, equally important, to minimize the volume of the cell stem which emerges from the coolant bath and is not at a uniform temperature.

15.4 EVACUATION AND DEGASSING

After attaching the sample cell containing sample to the apparatus, caution must be used to prevent powder from blowing out during the initial evacuation. If this occurs, it is necessary to recalibrate the various parts of the apparatus. A plug of glass wool inserted at the junction of the capillary tube and the wider part of the sample cell will prevent particles from entering

Table 15.1 Correction factors for nonideality (calculated from Emmett and Brunauer [125])

Gas	Temperature (°C)	α
N_2	−195.8	6.58×10^{-5}
	−183	3.78×10^{-5}
O_2	−183	4.17×10^{-5}
Ar	−195.8	11.4×10^{-5}
	−183	3.94×10^{-5}
CO	−183	3.42×10^{-5}
CH_4	−140	7.79×10^{-5}
NO	−140	5.26×10^{-5}
N_2O	−78	7.68×10^{-5}
CO_2	−78	2.75×10^{-5}
	25	1.559×10^{-5}
$n - C_4H_{10}$	0	14.2×10^{-5}
	25	4.21×10^{-5}

the manifold. The glass wool, however, will contribute to adsorption and represents a potential source of error.

During degassing, when the sample is being heated, a vacuum must be maintained in the rest of the apparatus to prevent condensation of vapor in the cooler parts of the equipment.

15.5 TEMPERATURE CONTROL

For maximum accuracy, the calibrated volumes and the manifold should be maintained at constant temperature by using a water or constant temperature air jacket. Alternatively, the temperature may be closely monitored and the volume adjusted accordingly. The liquid nitrogen level should be maintained as near constant as possible to prevent any variation in the effective volume of the sample cell.

15.6 ISOTHERMS

Because the shape of the isotherm establishes the quantity adsorbed at various relative pressures, it is often difficult to predict *a priori* the exact dosing quantity to admit to the calibrated volumes in order to obtain the desired equilibrium pressure. If the first dose of adsorbate equilibrates at a relative pressure higher than desired because of insufficient adsorption, it is best to reevacuate the sample and admit a smaller initial dose. Thereafter, smaller dose requirements can be anticipated for subsequent data points.

Conversely, if the first dose equilibrates at a lower relative pressure than desired, it is necessary only to dose a second time to attain a higher equilibrium pressure.

Because of the dosing technique associated with the vacuum volumetric method, there exists a potential source of error which in principle cannot be avoided on samples that are slow to equilibrate. Fig. 15.2 represents a plot of the pressure in the sample cell versus time. The pressure up to time t_1 represents the equilibrium pressure in the sample cell before a new quantity of adsorbate is admitted. At time t_1, adsorbate is admitted into the sample cell and is accompanied by a rapid pressure rise. The pressure then gradually decreases to a new equilibrium value at time t_2. If the decay to the new equilibrium pressure is slow, it is possible for open or accessible parts of the surface, such as the interior of wide pores, to contain more adsorbate before equilibrium is attained than after equilibrium is established. Less accessible regions, such as the interior of long narrow pores, will adsorb slowly and as the pressure falls because of adsorption in these pores, desorption must occur from the more accessible pores which tend to equilibrate more rapidly. If a porous sample is subjected to increasing pressure, followed by decreasing pressure, an unusual situation develops. This is because desorption cannot occur along the desorption isotherm and if the highest pressure reached is less than the saturated pressure, the quantity adsorbed will slide off the adsorption isotherm toward the adsorption isotherm. Thus, depending upon the final equilibrium pressure, the data point will lie somewhere between the two curves that comprise the hysteresis loop. It may well be that this effect occurs to some extent on any sample that contains pores with one or more constrictions. If the constriction is only slightly wider than the adsorbate dimensions, entrance into the wide inner pore will be hindered due to the strong interaction with the walls in the narrow portion. Adsorbate molecules will be able to pass through the constriction only if they possess sufficient thermal energy to

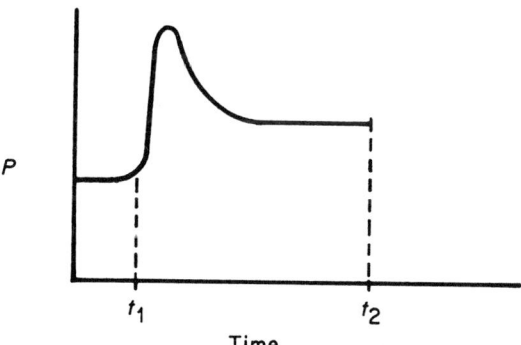

Figure 15.2 Adsorbate pressure before and after dosing, showing pressure overshoot before equilibrium is attained.

overcome the higher adsorption potential within this region. This situation is similar to the Arrhenius [126] concept of an activation energy used in chemical kinetics. According to Arrhenius, the specific reaction rate k is given by

$$k = Ae^{-E/RT} \tag{15.10}$$

where E is the energy that molecules must possess in order to react and A is related to the frequency with which molecules collide. By analogy, the rate at which molecules can penetrate through a constriction would be given by

$$\text{Rate} = Ae^{-E_{ad}/RT} \tag{15.11}$$

Here E_{ad} is the adsorption potential within the constriction. Equation (15.11) predicts that adsorption in pores with one or more constrictions will proceed more rapidly at elevated temperatures. Several examples of this type of 'activated physical adsorption' have been found [127–130].

For a surface containing area in pores which possess constrictions leading into wide inner bodies, one can approximate the ratio of the adsorption rates of the more accessible parts of the surface and the portion of the surface in the wide inner body. If the adsorption potential in a constriction is 2 kcal more than that on an open surface, the ratio of adsorption rates at liquid nitrogen temperature, assuming the same pre-exponential value A, will be

$$\frac{\text{Rate}_1}{\text{Rate}_2} = \frac{A_1}{A_2} \frac{e^{-0/2 \times 77}}{e^{-2000/2 \times 77}} = 4.3 \times 10^5 \tag{15.12}$$

assuming that the activation energy for adsorption on a readily accessible part of the surface is zero. Therefore, it may take weeks or months for equilibrium to be established in the pores with constrictions. A rate ratio of only two or three would be sufficient to cause adsorption and then desorption to occur from the accessible or wider pores leading to the data point residing in the hysteresis regions.

The consequences of this type of 'activated physical adsorption' is not only that the quantity adsorbed can lie off the isotherm but also that the measured quantity of adsorption is far less than the equilibrium value. No experiments have been conducted to illustrate whether or not the quantity adsorbed lies within the hysteresis loop. The occasional failure of the vacuum volumetric method to agree with the dynamic method, which is not subject to any pressure overshoot, may in part be attributed to this phenomena.

15.7 LOW SURFACE AREAS

By using small calibrated volumes, carefully measuring the void volume and maintaining constant temperature by thermostating the adsorption apparatus, surface areas as low as approximately $1.0\,\text{m}^2$ can be measured. The principal obstacle to measuring areas lower than this with nitrogen in a volumetric system lies in accurately measuring the void volume. On small areas, the quantity of adsorbate remaining in the void volume is large compared to the amount adsorbed and, indeed the void volume error can be larger than the volume adsorbed.

The number of molecules trapped in the void volume can be reduced by using adsorbates with low vapor pressures. Beebe, Beckwith, and Honig [131] were the first to use krypton for this purpose. Litvan [132] reported the vapor pressure of krypton at liquid nitrogen temperature as 2.63 torrs. Therefore, the amount of krypton remaining in the void volume, when monolayer coverage occurs, will be much less than nitrogen, whereas the amount of adsorption will be only slightly less by approximately the ratio of cross-sectional areas of nitrogen and krypton, or about 16.2/19.5.

To measure the low pressures associated with krypton adsorption, a thermocouple or thermistor can be used after calibration against a McLeod gauge.

Argon at liquid nitrogen temperature exhibits an equilibrium pressure of 187 torrs. It offers the advantage of a lower vapor pressure than nitrogen, which will reduce the void volume error while retaining ease of pressure measurements. However, the cross-sectional area of argon is not well established and appears to vary according to the surface on which it is adsorbed.

When using adsorbents with low vapor pressure, no correction for non-ideality is necessary. However, another type of correction is required due to 'thermal transpiration' [133] which occurs if two parts of a system contain the same gas at the same pressure but at different temperatures. Thermal transpiration can be understood by considering the number of collisions n per square centimeter per second which a gas makes with the walls of its container

$$n = \frac{\bar{N}P}{(2\pi \bar{M}RT)^{\frac{1}{2}}} \tag{15.13}$$

A tube connecting the two parts of a system will admit more molecules on the cold side than on the warm side. Consequently, some gas will flow from the cold to the warm ends of the tube until the pressure on the warm side increases sufficiently to permit the following equality:

$$n_{\text{cold}} = n_{\text{warm}} \tag{15.14}$$

so that

$$\frac{P_{\text{cold}}}{T^{\frac{1}{2}}_{\text{cold}}} = \frac{P_{\text{warm}}}{T^{\frac{1}{2}}_{\text{warm}}} \tag{15.15}$$

Then,

$$P_{warm} = \left(\frac{T_{warm}}{T_{cold}}\right)^{\frac{1}{2}} P_{cold} \tag{15.16}$$

For the effects of thermal transpiration to be observed, it is necessary that the mean free molecular path be greater than the tube diameter. At higher pressures and smaller mean free paths, molecular collisions destroy the effect.

In the pressure range useful for krypton adsorption, the mean free path is approximately equal to the diameter of the stem of the sample cell. Equation (15.16) is not applicable in this range and an empirical equation devised by Rosenberg [134] must be used. His equation is

$$P_c = P_w \left[1 - \frac{0.49}{37.2(dP_w)^2 + 14.45\,dP_w + 1}\right] \tag{15.17}$$

where P_c and P_w refer to the pressure in the sample cell at 77 K and room temperature, respectively and d is the inside diameter of the sample cell stem (mm).

15.8 SATURATED VAPOR PRESSURE, P_0 OF NITROGEN

The term P_0 is defined as the saturated equilibrium vapor pressure exhibited by pure nitrogen contained in the sample cell when immersed in liquid nitrogen coolant. The temperature of the liquid nitrogen, usually held in a Dewar flask, is always found to be somewhat different than the normal boiling point because of dissolved impurities from the atmosphere and because of ambient pressure fluctuations. Dissolved impurities usually increase the bath temperature sufficiently to cause the vapor pressure of pure liquid nitrogen within the sample cell to increase by 10–20 torrs above ambient pressure. Therefore, adding 15 torrs to ambient pressure will usually give the correct saturated pressure to within 5 torr.

When measuring surface areas, the effect of using an incorrect saturated pressure is further reduced by the nature of the BET plot since both the ordinate and abscissa will deviate in the same direction leaving the slope nearly constant. For BET C values near 100, a slope of 1000 and an intercept of 10, an error of 15 torrs in a total of 760 torrs or about 2% will produce less than 1% error in surface area.

When pore sizes are analyzed above the usual BET range of relative pressures, the error associated with saturated pressure measurements is insignificant when compared to the many other assumptions which are made.

When highly accurate P_0 measurements are necessary, a platinum resistance thermometer can be placed in the bath adjacent to the sample cell. Equation

(14.1) can then be used to calculate the nitrogen vapor pressure. Alternatively and perhaps preferably, the vapor pressure can be measured directly by condensing nitrogen in a cell contained in the coolant and connected directly to a manometer or sensitive pressure gauge.

15.9 AUTOMATED INSTRUMENTATION

Recently a somewhat novel approach was developed for the measurement of several (up to six) isotherms simultaneously and independently [135]. Fig. 15.3 shows six sample cells and six P_0 stations connected to a common manifold. In this system, each sample cell can be isolated from the manifold by a valve; a pressure transducer is connected to each cell and one is also attached to the manifold.

Prior to each analysis, during initialization, the computer-controlled instrument admits adsorbate to the manifold with the valves to the sample cells open. The computer checks the output of each cell transducer to confirm that they agree with the primary pressure transducer mounted on the manifold by monitoring the pressure at ten different values. Any transducer found to be misaligned is automatically recalibrated. After the void volume in each cell is determined by dosing with helium, the first cell is evacuated and then dosed with nitrogen gas. The dosing algorithm requires that the manifold be brought to a pressure such that, when the valve to the cell is opened the cell pressure will reach a pre-set target pressure, assuming no adsorption occurs. Once the target pressure is achieved the cell is isolated from the manifold and the approach to equilibrium is monitored by the cell transducer. In this manner, the manifold is now free to dose another cell while the first one is reaching its equilibrium pressure. If a cell does not reach its target equilibrium pressure after it is dosed, as sensed by the cell transducer, that cell will be redosed automatically as often as necessary until it reaches the target pressure. While waiting for equilibrium, after dosing a cell, the manifold is employed to redose other cells whose target pressure was not reached. In this manner, six adsorption and desorption isotherms can be obtained in only slightly more time than that required for one isotherm.

An interesting feature of this new instrumentation is the ability to set the target pressure tolerance and the time allowed to reach the target pressure. Thus, with a tight tolerance and long equilibration times, the instrument will produce data at almost exactly the requested relative pressures while, with looser tolerances and shorter equilibration times, the analysis can proceed more rapidly and produce data points with less precision, but no less accuracy.

15.10 QUASI-EQUILIBRIUM SYSTEMS

In recent years, there have been a number of reports of quasi-equilibrium systems for the determination of sorption isotherms [136]. These systems

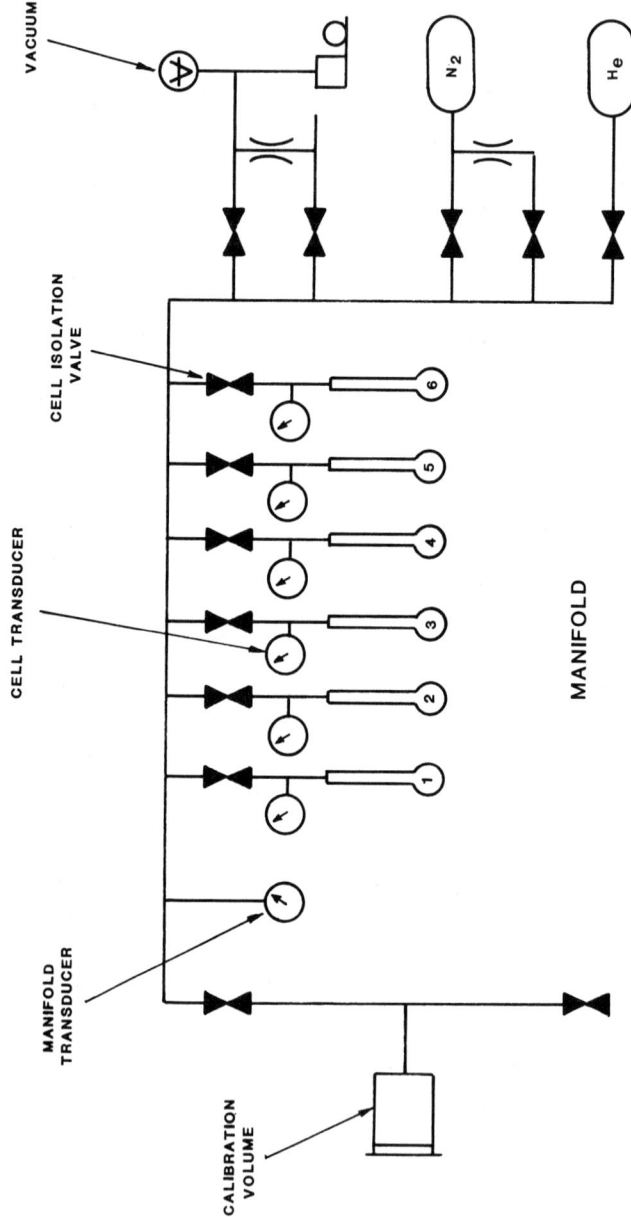

Figure 15.3 Schematic of automated sorption system.

involve a constant bleed of adsorbate gas into the sample cell which results in a continuous scan of the adsorption isotherm and, similarly, a constantly decreasing pressure to produce a scan of the desorption isotherm. The quantity of gas adsorbed in the sample is calculated by integrating the amount of gas admitted into the cell at various pressures. The difference between the gas admitted and the gas contained in the void volume represents the amount adsorbed.

Although the quasi-equilibrium method is rapid, it often produces data with lower uptake of gas on the adsorption isotherm [136] and insufficient desorption on the desorption curve. In the region of capillary condensation, longer times are required to achieve equilibrium between the gas and capillary condensed phases than the equipment can provide. To confirm that equilibrium has been established in the quasi-equilibrium method the analysis should be re-run with slower gas bleed rates. The validity of the analysis is confirmed if identical data is obtained at two different gas flows. Attempts by the authors to construct accurate isotherms using this method indicated that inaccurate data is often obtained in the region of capillary condensation, as well as at low pressures on the desorption isotherm where evaporation out of small pores and from highly energetic sites causes equilibrium to be established very slowly.

16
Dynamic methods

16.1 INFLUENCE OF HELIUM

In 1951 Loebenstein and Deitz [137] described an innovative gas adsorption technique which did not require the use of vacuum. They adsorbed nitrogen out of a mixture of nitrogen and helium which was passed back and forth over the sample between two burettes by raising and lowering attached mercury columns. Equilibrium was established by noting no further change in pressure with additional cycles. The quantity adsorbed was determined by the pressure decrease at constant volume. Successive data points were acquired by adding more nitrogen to the system.

The results obtained by Loebenstein and Deitz agreed with vacuum volumetric measurements on a large variety of samples with a wide range of surface areas. They were also able to establish that the quantities of nitrogen adsorbed were independent of the presence of helium.

Because all the methods mentioned in this chapter require that the adsorbate be mixed with an inert carrier gas, usually helium, some discussion regarding the possible influence of helium is necessary.

The forces leading to liquefaction and adsorption are the same in origin and magnitude. Gases substantially above their critical temperature, T_c, cannot be liquefied because their thermal energy is sufficient to overcome their intermolecular potential. Although the adsorption potential of a gas can be greater than the intermoleular potential, helium is, nevertheless, not adsorbed at liquid nitrogen temperature (77 K) because this temperature is still more than 14 times the critical temperature of helium (5.3 K).

Furthermore, in order to be considered adsorbed, a molecule must reside on the surface for a time τ at least as long as one vibrational cycle of the adsorbate normal to the surface. The time for one vibration is usually of the order 10^{-13} seconds which by equation (16.1) makes τ about 2×10^{-13} seconds at 77 K, (see equation 4.4 and the following paragraph).

$$\tau = 10^{-13} e^{100/RT} = 1.91 \times 10^{-13} \, \text{s} \tag{16.1}$$

The value of 100 cal mol^{-1} chosen for the adsorption energy of helium is consistent with the fact that helium has no dipole or quadrapole and is only slightly polarizable. Thus, it will minimally interact with any surface. Based upon reflections of a helium beam from LiF and NaCl cleaved surfaces, de Boer [138] estimated the adsorption energy to be less than 100 cal mol^{-1}.

At 77 K the velocity of a helium atom is 638 m s^{-1}, so that in 1.91×10^{-13} s it will travel $1.91 \times 10^{-13} \times 638 \times 10^{10}$ or 1.2 Å. Thus, the condition that the adsorbate reside near the surface for one vibrational cycle is fulfilled by the normal velocity of helium and not by virtue of being adsorbed. Stated in alternate terms, the density of helium near a solid surface at 77 K is independent of the surface and is the same as the density remote from the surface.

Molecular collisions with the adsorbed film by helium will certainly be no more destructive than collisions made by the adsorbate. In fact, helium collisions will be less disruptive of the adsorbed film structure since the velocity of helium is, on the average, 2.6 times greater than that of nitrogen at the same temperature, while a nitrogen molecule is 7 times heavier. Thus, the momentum exchange due to nitrogen collisions will be the more disruptive.

The thermal energy of helium at 77 K is about 220 cal mol^{-1}. The heat of vaporization of nitrogen at 77 K is 1335 cal mol^{-1}, which may be taken as the minimum heat of adsorption. A complete exchange of thermal energy during collisions between a helium atom and an adsorbed nitrogen molecule would not be sufficient to cause desorption of the nitrogen.

A further point to be made is that the extent of helium's influence, if any, on the accuracy of an adsorption measurement, as a carrier gas in a dynamic system must be weighed against its effect when used to calibrate the void volume in a vacuum system.

16.2 NELSON AND EGGERTSEN CONTINUOUS FLOW METHOD

Nelson and Eggertsen [139], in 1958, extended the Loebenstein and Dietz technique by continuously flowing a mixture of helium and nitrogen through the powder bed. They used a hot wire thermal conductivity detector to sense the change in effluent gas composition during adsorption and desorption, when the sample cell was immersed into and removed from the bath, respectively.

Fig. 16.1 illustrates a simplified continuous flow apparatus. Fig. 16.2 is a schematic of the flow path arrangement using a four-filament thermal conductivity bridge.

In Fig. 16.1, a mixture of adsorbate and carrier gas of known concentration is admitted into the apparatus at a. Valve V_1 is used to control the flow rate. When the system has been purged, the detectors are zeroed by balancing the bridge (see Fig. 16.8). When the sample cell b is immersed in the coolant, adsorption commences and detector D_2 senses the increased helium concentration. Upon completion of adsorption, D_2 again detects the same

176 Dynamic methods

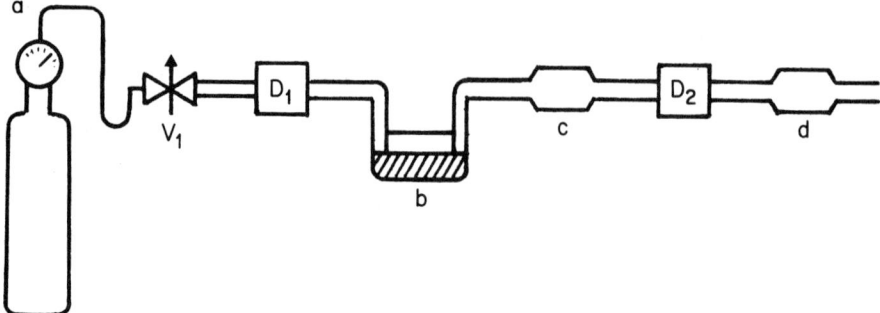

Figure 16.1 Simplified continuous flow apparatus.

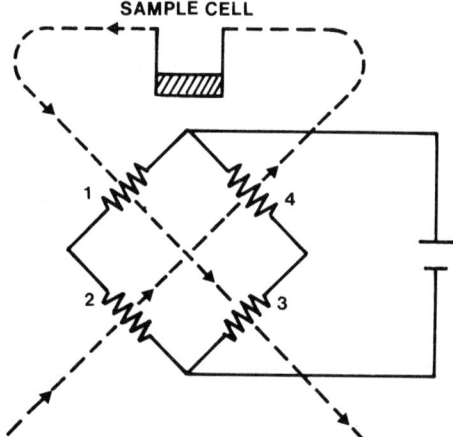

Figure 16.2 Flow path using a four-filament bridge. Filaments 1 and 3 are D_2, 2 and 4 are D_1. Dotted line represents gas flow.

concentration as D_1 and the signal returns to zero. When the coolant is removed, desorption occurs as the sample warms and detector D_2 senses the increased nitrogen concentration. Upon completion of desorption, the detectors again sense the same concentration and the signal returns to its initial zero value. Volume c is a wide segment of tubing used to decrease the linear flow velocity of the gas ensuring its return to ambient temperature prior to entering D_2. Volume d is also a wide segment of tubing which acts as a ballast in order to prevent air from being sucked into D_2 as the cell is cooled and the gas contracts. Fig. 16.3 illustrates the detector signals due to adsorption and subsequent desorption.

Figs. 16.4 and 16.5 illustrate a parallel flow arrangement which has the

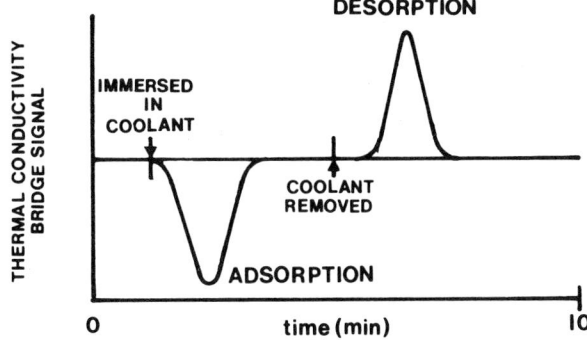

Figure 16.3 Adsorption and desorption peaks in a continuous flow apparatus.

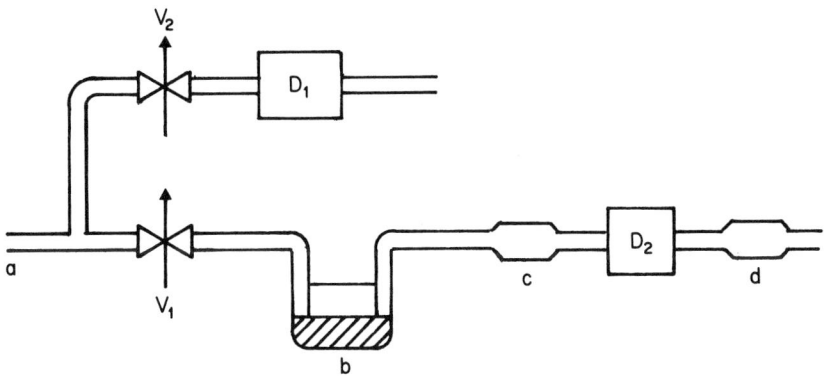

Figure 16.4 Parallel flow circuit.

advantage of requiring shorter purge times when changing gas composition but is somewhat more wasteful of the mixed gases. The symbols shown in Fig. 16.4 have the same meaning as those used in Fig. 16.1.

16.3 CARRIER GAS AND DETECTOR SENSITIVITY

To be effective, the carrier gas must fulfill two requirements. First, it cannot be adsorbed at the coolant temperature; second, it must possess a thermal conductivity sufficiently different from that of the adsorbate that small concentration changes can be detected.

178 Dynamic methods

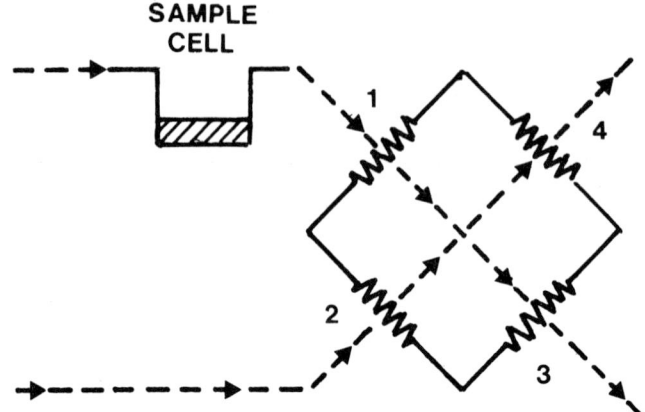

Figure 16.5 Parallel flow path using a four-filament bridge. Filaments 1 and 3 are D_2, 2 and 4 are D_1. Dotted lines are gas flow paths.

To understand the effect of the carrier gas on the response of the thermal conductivity detector, consider the steady state condition that prevails when the resistive heat generated in the hot wire filament is exactly balanced by the heat conducted away by the gas. This condition is described by equation (16.2)

$$i^2 R = ck(t_f - t_w) \qquad (16.2)$$

where i is the filament current, R is the filament resistance, k is the thermal conductivity of the gas mixture, and t_f and t_w are the filament and the wall temperatures, respectively. The constant c is a cell constant that reflects the cell geometry and the separation of the filament from the wall which acts as the heat sink.

When the gas composition is altered due to adsorption or desorption, the value of k changes by Δk which in turn alters the filament temperature by Δt_f. Under the new conditions equation (16.2) can be rewritten as

$$i^2 R = c(k + \Delta k)(t_f + \Delta t_f - t_w) \qquad (16.3)$$

Equating the right hand side of equations (16.2) and (16.3) gives

$$k(t_f - t_w) = (k + \Delta k)(t_f + \Delta t_f - t_w) \qquad (16.4)$$

By neglecting the term $\Delta k \Delta t_f$, a second order effect, equation (16.4) rearranges to

Carrier gas and detector sensitivity

$$\Delta t_f = \frac{\Delta k}{k}(t_w - t_f) \tag{16.5}$$

The change in filament resistance ΔR is directly proportional to the small temperature change Δt_f and is given by

$$\Delta R = \alpha R \Delta t_f \tag{16.6}$$

where α is the temperature coefficient of the filament, dependent on its composition and R is the filament resistance at temperature t_f; thus

$$\Delta R = \alpha R \frac{\Delta k}{k}(t_w - t_f) \tag{16.7}$$

Equation (16.7) requires that Δk be as large as possible for maximum response under a fixed set of operating conditions.

A fortunate set of circumstances leads to a situation in which the same molecular properties that impart minimal interactions or adsorption potentials also lead to the highest thermal conductivities. Molecules of large mass and many degrees of vibrational and rotational freedom tend to be more polarizable and possess dipoles and quadrapoles which give them higher boiling points and stronger interactions with surfaces. These same properties tend to reduce their effectiveness as thermal conductors.

Helium possesses only three degrees of translational freedom and hydrogen the same, plus two rotational and one vibrational degree. However, because of hydrogen's light weight, it has the highest thermal conductivity of all gases, followed by helium. Either of these two gases fulfills the requirement for adequately high thermal conductivities so that Δk in equation (16.7) will be sufficiently large to give good sensitivity with any adsorbate. Helium, however, is usually used in continuous flow analysis because of the hazards associated with hydrogen.

Fig. 16.6 is a plot of the thermal conductivity of mixtures of helium and nitrogen obtained on an apparatus similar to that described in the next section. Characteristically, the thermal conductivity of most mixtures does not vary linearly with concentration. The slope of the curve at any point determines the value of Δk and, therefore, the detector response. Fig. 16.6 also illustrates that the greater the difference between thermal conductivities of the adsorbate and carrier gas, the higher will be the slope and therefore the detector response.

The shape of the curve shown in Fig. 16.6 is fortuitous in as far as the continuous flow method is concerned. For reasons to be discussed later, the desorption signal (see Fig. 16.3) is generally used to calculate the adsorbed volume. When, for example, $1.0 \, cm^3$ of nitrogen is desorbed into $9.0 \, cm^3$ of helium, the concentration change is 10%. However, when $1.0\,\text{cm}^3$ of nitrogen is desorbed into $9.0\,\text{cm}^3$ of a 90% nitrogen-in-helium mixture, the absolute change is only 1%. Therefore, the increase in slope at high nitrogen concentrations enables smaller concentration changes to be detected when data at high relative pressures is required.

Fig. 16.6 was prepared by flowing helium through one detector while varying the helium to nitrogen concentration ratio through the second detector.

16.4 DESIGN PARAMETERS FOR CONTINUOUS FLOW APPARATUS

The thermal conductivity (T.C.) detector consists of four filaments embedded in a stainless steel or brass block which acts as a heat sink. The T.C. detector is extremely sensitive to temperature changes and should be insulated to prevent temperature excursions during the time in which it takes to complete an adsorption or desorption measurement. Long-term thermal drift is not significant because of the calibration procedure discussed in the next section and therefore, thermostating is not required.

Fig. 16.7 shows a cross-sectional view of a T.C. block and the arrangement of the filaments relative to the flow path. The filaments shown are electrically connected, external to the block, and constitute one of the two detectors.

The filaments must be removed from the flow path, unlike the conventional 'flow over' type used in gas chromatography, because of the extreme flow variations encountered when the sample cell is cooled and subsequently warmed. Flow variations alter the steady state heat transport from the filaments, leaving them inadequate time to recover before the concentration

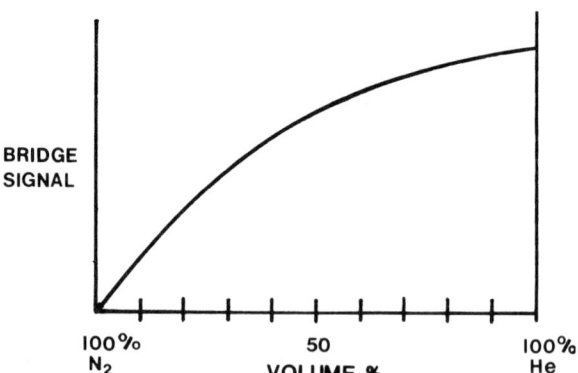

Figure 16.6 Thermal conductivity bridge unbalance. He through D_1 and He–N_2 mixture through D_2.

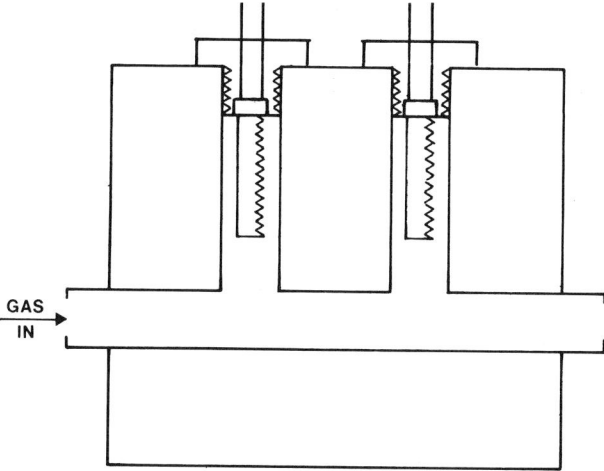

Figure 16.7 Thermal conductivity block with filaments located out of the flow path.

change from adsorption or desorption is swept into the detector. When this occurs the base line from which the signal is measured will be unstable. By removing the filaments from the flow path and allowing diffusion to produce the signal, the problem of perturbing the filaments is completely solved. However, the trade off is nonlinear response characteristics. Since the thermal conductivity of the gas mixture is not linear with concentration, this additional nonlinearity poses no additional problems and is accommodated by calibration of signals which is the topic of the next section.

A suitable electronic circuit to provide power to the filaments, a means of zeroing or balancing the T.C. bridge, adjusting the filament current, attenuating the signal and adjusting the polarity is shown in Fig. 16.8. Signals produced by adsorption or desorption can be fed to a strip chart recorder for a continuous record of the process and to a digital integrator for summing the area under the adsorption and desorption curves.

Fig. 16.9 is the flow schematic for a commercial continuous flow apparatus. Four gas concentrations from premixed tanks can be selected by valve V_s. Alternatively, gases can be blended internally by controlling their flows with needle valves N_1 and N_2. A third choice is to feed mixtures to the apparatus from two linear mass flow controllers. Flow meters M_1 and M_2, operating under pressure to extend their range, indicate input flow rates. Bubble flow meter B is used to calibrate the flow meters; however, the bubble meter need not be calibrated since it is used to establish flow ratios when blending gases internally. Gauge, $G_1(0 - 0.4 \text{ MPa})$, indicates the input pressure under which the flow meters operate. Needle valves N_1 and N_2 control the input flow rates. Cold trap T removes contaminants from the gas stream. After leaving the

182 Dynamic methods

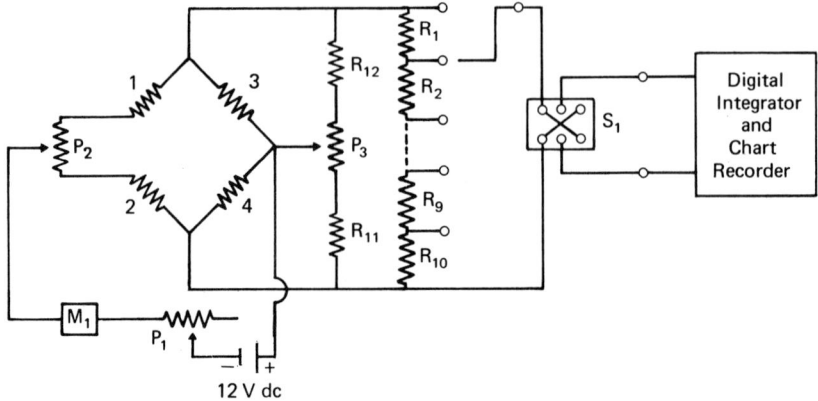

Figure 16.8 Thermal conductivity bridge electronic circuit. 12 V dc power supply, stable to 1 mV. Ripple is not significant due to thermal lag of the filaments. P_1, 100 ohms for filament current control; M_1, milliam meter, 0–250 mA; P_2, 2 ohms for coarse zero. Filaments 1 and 4 are detector 1; 2 and 3 are detector 2. P_3, 1 ohm for fine zero; R_{11}, R_{12}, padding resistors ~64 ohms; R_1–R_{10}, attenuator resistors 1, 2, 4 ... 512 ohms; S_1 (D.P.D.T.) switch for polarity. Attenuator resistors are $\frac{1}{4}$% ww, lowest temperature coefficient; all others are 1%.

cold trap but before flowing into detector D_1, the gas passes through wide diameter tubing E_1 which reduces the linear flow velocity and provides time for warming to ambient temperature.

Upon leaving D_1 the flow is split, the larger flow goes to sample cell S_1, then through filter F, a thermal equilibrating tube E_2 and a flow meter M_3 which indicates the flow through the sample cell. The smaller flow, controlled by needle valve N_1 merges with the sample cell effluent before entering detector D_2. The length of wide-bore tubing E_3 serves as a ballast to prevent air from entering D_2 when the sample is immersed in the coolant. For large quantities of adsorbed gas a long path can be employed by opening valve V_{10}. This prevents the gas from reaching the detector before the flow has returned to its original rate.

Splitting the flow serves as a means of diluting the adsorption and desorption concentration peaks in order continuously to adjust the signal height using needle valve N_3. The electronic attenuator shown in Fig. 16.8 adjusts the signal by factors of two for each step.

Valves N_4 and V_5 allow a slow bleed of adsorbate to the 'out' septum and then to a sample cell positioned at the degassing station. The slow flow of gas through this part of the circuit provides a source of adsorbate for the purpose of signal calibration and as a purge for the degassing station holding cell S_2.

The septum labelled 'in' is used to inject known quantities of adsorbate into the flow to simulate a desorption signal for the purpose of calibration.

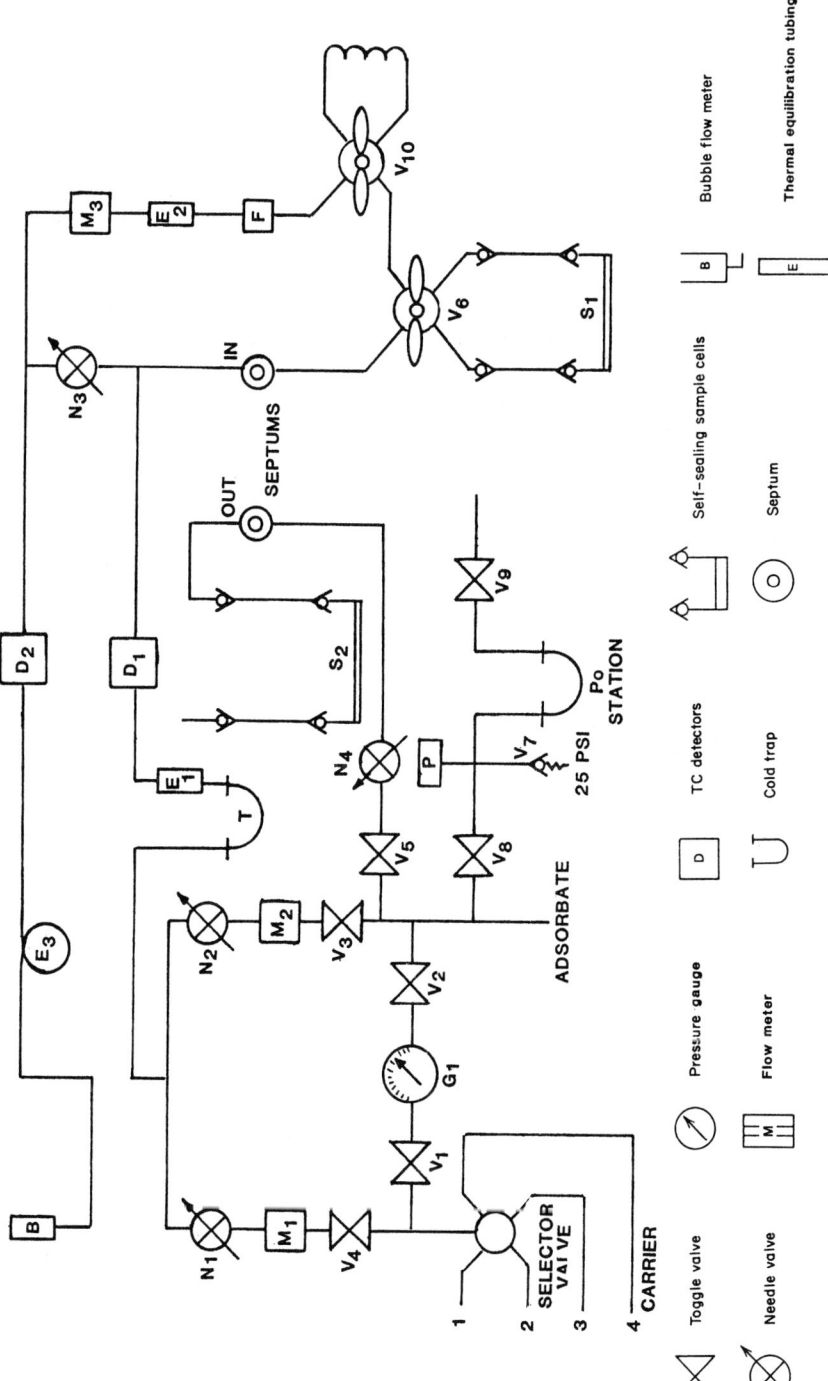

Figure 16.9 Flow diagram for continuous flow system. Courtesy of Quantachrome Corporation.

184 Dynamic methods

In order to measure the saturated vapor pressure pure adsorbate is admitted through V_8 to the P_0 station while it is immersed in liquid nitrogen. The pressure is measured by transducer P.

Valve V_6 is rotated to maintain flow continuity through the system when the sample cell is removed.

In order to avoid contamination of the degassed sample when transferring from the degassing to the analysis position, the cells are mounted in spring-loaded self-sealing holders that close when disconnected and open when placed in position.

Sample cells consist of a wide variety of designs for various applications. Fig. 16.10 illustrates seven cells used for various types of samples. Their specific applications and limitations are given in more detail in Section 16.7. Each of the cells shown is made of Pyrex glass. They are easily filled and cleaned. The cells range from four to five inches in length with stem inside and outside diameters of 0.15 and 0.24 inches, respectively.

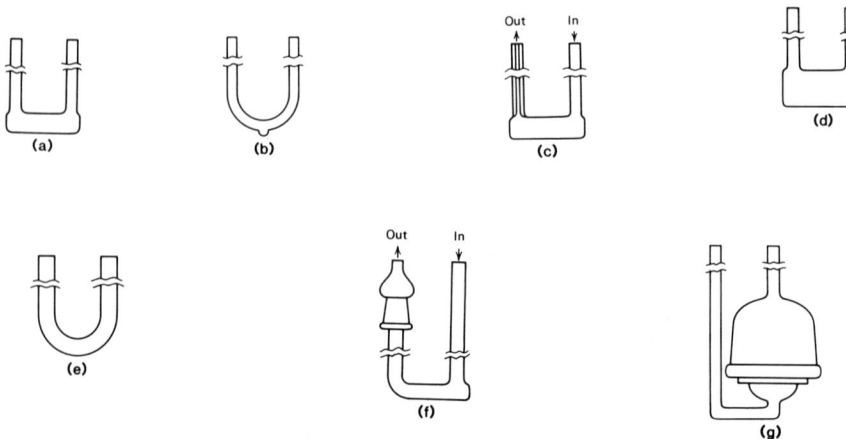

Figure 16.10 Sample cell designs. (a) Conventional sample cell — for most powder samples with surface areas greater than $0.2\,m^2$ in the cell. For samples less than $0.2\,m^2$, this cell can be used with krypton as the adsorbate. (b) Micro cell — used for very high surface area samples or for low area samples that exhibit thermal diffusion signals. Because of the small capacity of the micro cell, low area samples must be run on high sensitivity settings. (c) Capillary cell — useful for minimizing thermal diffusion when a large quantity of low area sample is required. (d) Macro cell — used with krypton when a large quantity of sample is required. Also used for chemisorption when total uptake is small. (e) Large U-tube cell — for larger particles or bulk samples of high area with nitrogen or low area with krypton. (f) Pellet cell — used for pellets or tablets. High surface area with nitrogen or low area samples with krypton. (g) Monolith catalyst cell — for monolithic catalysts and other samples of wide diameter that must be measured as one piece.

16.5 SIGNALS AND SIGNAL CALIBRATION

The signal intensity created by an adsorption or desorption peak passing through the detector is dependent upon the attenuator setting, the filament current, and the design of the T.C. detector. Also, as stated previously, the detector response is nonlinear. These circumstances require that the adsorption or desorption signals be calibrated by introducing a volume of carrier or adsorbate gas into the flow stream. An expeditious and accurate method of calibration is the withdrawal of a sample of adsorbate from the 'out' septum (see Fig. 16.9) with a precision gas syringe and the injection of a known volume into the flow stream through the 'in' septum.

Usually the desorption peak is calibrated because it is free of tailing. By immersing the cell in a beaker of water immediately after removal from the liquid nitrogen, the rate of desorption is hastened. Heat transfer from the water is more rapid than from the air, therefore, a sharp desorption peak is generated. The calibration signal should be within 20% of the desorption signal height in order to reduce detector nonlinearity to a negligible effect.

The area under the desorption peak A_d and the area under the calibration peak A_c are used to calculate the volume desorbed V_d from the sample according to equation (16.8)

$$V_d = \frac{A_d}{A_c} V_c \tag{16.8}$$

where V_c is the volume of adsorbate injected. Signals can be plotted on a chart recorder and their areas can be obtained by commercial digital integrators of the type used in gas chromatography (see Fig. 16.8). Alternatively, the areas can be measured by planimetry, counting boxes, or by cutting and weighing. The latter methods, however, are less accurate and slower than modern digital integrators.

Equation (16.8) requires no correction for ideality since the volume desorbed as measured at ambient temperature and pressure. Because desorption occurs at room temperature, it is complete and represents exactly the quantity adsorbed. For vapors adsorbed near room temperature, the sample can be heated to ensure complete desorption.

A detailed analysis of the chart record, shown in Fig. 16.11, corresponding to a complete adsorption, desorption, calibration and concentration change cycle discloses that at point a the sample cell is immersed in the coolant which produces the adsorption peak P_1. At b, the polarity is reversed and the base line is reestablished. Point c represents removal of the coolant bath which leads to the desorption peak P_2. The calibration peak P_3 results from the calibration injection made at point d. At e, a new gas concentration is admitted into the apparatus, which produces a steady base line at f where the detector is rezeroed and the cycle is repeated. The total time for a cycle is

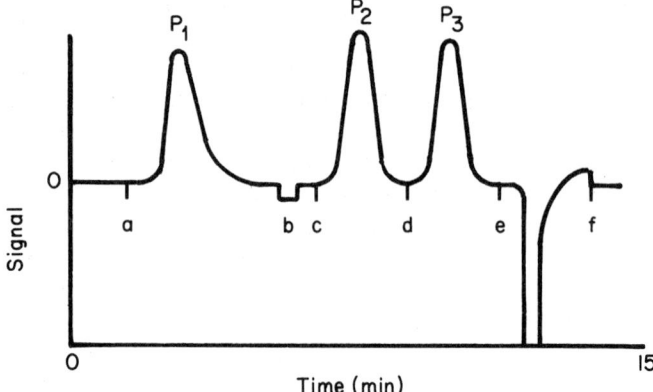

Figure 16.11 Complete cycle for one data point.

usually about 15 minutes. Some time savings can be gained by combining the purge step (e–f) with the adsorption step (a–b). Familiarity with the apparatus usually allows the operator to choose the correct volume for calibration at particular attenuator and filament current settings or, alternatively, a calibration table can be prepared. If speed is essential the flow rate can be increased to hasten the cycle. However, this is at the risk of warming the adsorbent if excessive flow rates are used. Flow rates of $12-15\,cm^3\,min^{-1}$ allow ample time for the gas to equilibrate thermally before reaching the sample powder with all of the cells shown in Fig. 16.10. The presence of helium, with its high thermal conductivity, ensures rapid thermal equilibrium. Therefore, immersion depth of the cell is not critical provided that about two and one-half inches is in the coolant at the flow rates given previously.

Sample cells are not completely filled with powder; room is left above the surface for the unimpeded flow of gas. Although the gas flows over the powder bed and not through it, lower flow rates aid in ensuring against elutriation.

The areas under the adsorption and desorption peaks are usually not exactly the same. This is due to the changing slope of Fig. 16.6. Adsorption produces concentration changes to the right or in the direction of decreased sensitivity, while desorption produces signals in the direction of increased sensitivity.

If calibration of the adsorption signal is desired, it is necessary to inject a known quantity of helium. The amount of helium used to calibrate the adsorption signal will usually vary considerably from the amount of nitrogen required for the desorption calibration. This situation arises because, for example, if $1\,cm^3$ of nitrogen is adsorbed out of a 10% flowing mixture, it will produce the equivalent of $9.0\,cm^3$ of helium. Therefore, calibration of the adsorption signal will require nine times more helium than the corresponding volume of nitrogen needed to calibrate the desorption signal. If C_{N_2} and C_{He} are the concentrations of nitrogen and helium in the flow stream and if V_{He} is the volume of helium used for calibration, then the volume of nitrogen adsorbed V_{ads} is given by

$$V_{ads} = V_{He} \left(\frac{C_{N_2}}{C_{He}}\right)\left(\frac{A_{ads}}{A_{cal}}\right) \tag{16.9}$$

where A_{ads} and A_{cal} are the areas under the adsorption and calibration signals, respectively.

When small signals are generated, it is difficult to make accurate injections of the required small amounts of gas. Karp and Lowell [140] have offered a solution to this problem which involves the injection of larger volumes of adsorbate diluted with the carrier gas. When a volume containing a mixture of nitrogen and helium V_{mix} is injected into the flow stream, the equivalent volume of pure nitrogen V_{N_2} is given by

$$V_{N_2} = V_{mix}\left(\frac{X_{N_2} - X'_{N_2}}{X'_{He}}\right) \tag{16.10}$$

where X_{N_2} is the mole fraction of nitrogen in the calibration mixture and X'_{N_2} and X'_{He} are the mole fractions of nitrogen and helium in the flow stream.

The signals must propagate through the system at identical flow rates. Calibration at a flow rate other than the flow rate associated with the adsorption or desorption peaks can lead to serious errors because the width of the peaks and therefore the peak areas are directly proportional to the flow rate. A good two-stage pressure regulator and needle valve provide adequately constant flow rates over the short time required for desorption and calibration.

Precision gas calibrating syringes can be obtained in various size ranges with no more than 1% volumetric error. Constant stroke adapters provide a high degree of reproducibility. Often, in the BET range of relative pressures, the calibration volumes remain nearly constant because the increased volume adsorbed at higher relative pressures tends to be offset by the decrease in the detector sensitivity. Thus, the same syringe may be used for a wide range of calibrations, which results in the syringe error not effecting the BET slope and only slightly altering the intercept, which usually makes a small contribution to the surface area. A syringe error of 1% will produce an error in surface area far less than 1% for those BET plots with a slope greater than the intercept or for high C values.

16.6 ADSORPTION AND DESORPTION ISOTHERMS BY CONTINUOUS FLOW

To construct the adsorption isotherm, the adsorption, desorption, and calibration cycle shown in Fig. 16.11 is repeated for each data point required. Errors are not cumulative since each point is independently determined. Relative pressures corresponding to each data point are established by measuring the saturated vapor pressure using any of the preceding methods or by adding 15 torrs to ambient pressure. Thus, if X is the mole fraction of adsorbent in the flow stream, the relative pressure is given by

$$\frac{P}{P_0} = \frac{XP_a}{P_a + 15} \tag{16.11}$$

where P_a is ambient pressure in torrs. At the recommended flow rates of 12–15 cm^3 min^{-1}, the flow impedance of the tubing does not raise the pressure in the sample cell.

The method used to construct the adsorption isotherm cannot be used to build the desorption isotherm. This is true because each data point on the adsorption curve reflects the amount adsorbed by a surface initially free of adsorbate. The desorption isotherm, however, must consist of data points indicating the amount desorbed from a surface that was previously saturated with adsorbate and subsequently equilibrated with adsorbate of the desired relative pressure. Karp, Lowell and Mustacciuolo [141] have demonstrated that the desorption isotherm and hysteresis loop scans can be made in the following manner. First, the sample is exposed, while immersed in the coolant, to a flow of pure adsorbate. The flow is then changed to the desired concentration, leading to some desorption until the surface again equilibrates with the new concentration. The coolant is then removed and the resulting desorption signal is calibrated to give the volume adsorbed on the desorption isotherm. The above procedure is repeated for each data point required, always starting with a surface first saturated with pure adsorbate.

To scan the hysteresis loop from the adsorption to the desorption isotherm, the sample, immersed in the coolant, is equilibrated with a gas mixture with a relative pressure corresponding to the start of the scan on the desorption isotherm. The adsorbate concentration is then reduced to a value corresponding to a relative pressure between the adsorption and desorption isotherms. When equilibrium is reached, as indicated by a constant detector signal, the coolant is removed and the resulting desorption signal is calibrated. Repetition of this procedure, each time using a slightly different relative pressure between the adsorption and desorption isotherms, yields a hysteresis scan from the adsorption to the desorption isotherm.

To scan from the desorption to the adsorption branch, pure adsorbate is first adsorbed, then the adsorbate concentration is reduced to a value giving a relative pressure corresponding to the start of the scan on the desorption isotherm. When equilibrium is established, as indicated by a constant base line, the adsorbate concentration is increased to give a relative pressure between the desorption and adsorption isotherms. After equilibrium is again established, the coolant is removed and the resulting signal is calibrated to yield a data point between the desorption and adsorption isotherms. This procedure repeated, each time using a different final relative pressure, will yield a hysteresis loop scan from the desorption to the adsorption isotherm.

Figs. 16.12 and 16.13 illustrate the results obtained using the above method on a porous amorphous alumina sample. A distinct advantage of the flow system for these measurements is that data points can be obtained where they are desired and not where they happen to occur after dosing, as in the vacuum volumetric method. In addition, desorption isotherms and hysteresis scans are accomplished with no error accumulation, void volume, or ideality corrections.

16.7 LOW SURFACE AREA MEASUREMENTS

The thermal conductivity bridge and flow circuits shown in Figs. 16.8 and 16.9 are capable of producing a full-scale signal (1.0 mV) when 0.01 cm³ of nitrogen is desorbed into a 30% nitrogen and helium mixture. To achieve stable operating conditions at this sensitivity, the thermal conductivity block

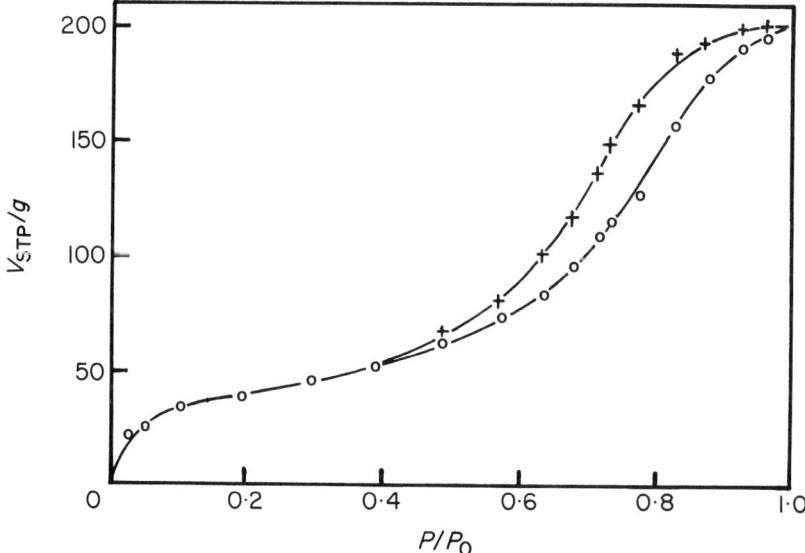

Figure 16.12 Adsorption and desorption isotherms of N_2 for 0.106 g sample of alumina. Adsorption, ○; desorption, +. V in cm³.

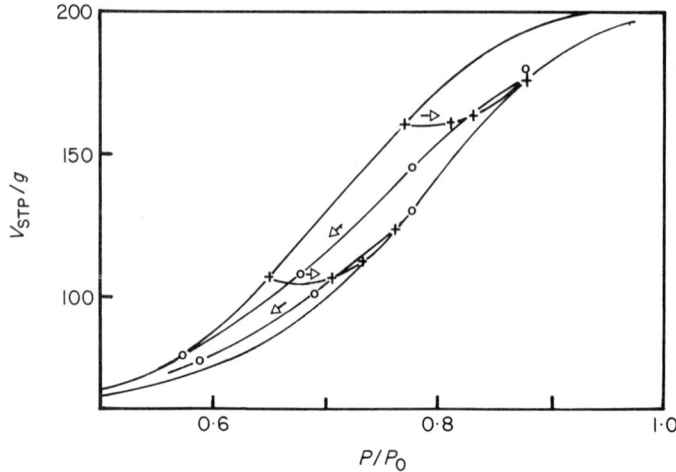

Figure 16.13 Hysteresis loop scan for same sample as Fig. 16.12. Desorption to adsorption, +; adsorption to desorption, ○.

will require about one hour to equilibrate thermally and the system must be purged of any contaminants.

A desorbed volume of $0.001 \, \text{cm}^3$, using nitrogen as the adsorbate, will correspond to about $0.0028 \, \text{m}^2$ ($28 \, \text{cm}^2$) of surface area if a single adsorbed layer were formed. An equivalent statement is that $0.0028 \, \text{m}^2$ is the surface area, measured by the single point method, on a sample which gives a high C value, if $0.001 \, \text{cm}^3$ were desorbed. Assuming that a signal 20% of full scale is sufficient to give reasonable accuracy for integration, then the lower limit for surface area measurement using hot wire detection is about $0.0006 \, \text{m}^2$ or $6 \, \text{cm}^2$. With thermistor detectors the lower limit would be still smaller.

Long before these extremely small areas can be measured with nitrogen, the phenomenon of thermal diffusion obscures the signals and imposes a higher lower limit [142]. Thermal diffusion produces a tendency for a gas mixture to separate when exposed to a changing temperature gradient. The arms of a sample cell where they enter the coolant experience a steep temperature gradient which changes for a short time interval until a steady state condition is attained. The resultant effect is for the heavier molecules to migrate to the high temperature region. In a static mixture of gases, the amount of thermal diffusion is a function of the time rate of change of the temperature gradient, the gas concentration, and the difference in masses of the molecules. In a flowing gas mixture, in the presence of adsorption, it is difficult to assess the exact amount of thermal diffusion.

Lowell and Karp [143] measured the effect of thermal diffusion on surface areas using the continuous flow method. Fig. 16.14 illustrates a fully developed anomalous desorption signal caused by thermal diffusion.

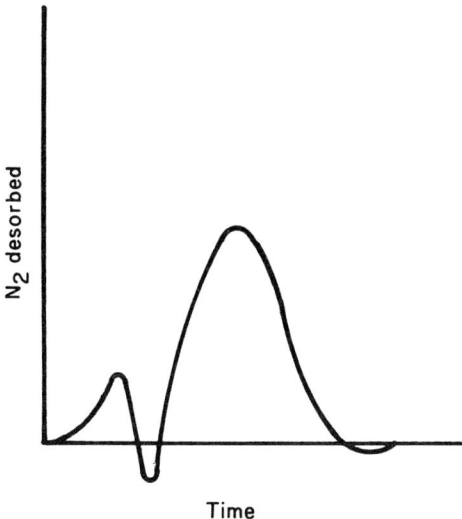

Figure 16.14 Desorption of a small volume of nitrogen.

As a result of the positive and negative nature of the signal, accurate integration of the true desorption peak is not possible. Table 16.1 shows the results of measuring the surface area of various quantities of zinc oxide using a conventional sample cell, Fig. 16.10(a).

When the same sample was analyzed using a micro cell, Fig. 16.10(b), the results obtained were considerably improved, as shown in Table 16.2.

The onset of thermal diffusion depends on the gas concentrations, the sample surface area, the rate at which the sample cools to bath temperature, and the packing efficiency of the powder. In many instances, using a conventional sample cell, surface areas less than $0.1 \, m^2$ can be accurately measured on well-packed samples that exhibit small interparticle void volume. The use of the micro cell (Fig. 16.10(b)) is predicated on the latter of these observations. Presumably, by decreasing the available volume into which the lighter gas can settle, the effects of thermal diffusion can be minimized. Although small sample quantities are used with a micro cell, thermal conductivity detectors are sufficiently sensitive to give ample signal.

Another cell design that aids in minimizing the effects of thermal diffusion is the capillary cell, Fig. 16.10(c). By using capillary tubing on the vent side of the cell, a sufficiently high linear flow velocity is maintained to prevent that arm from contributing to the problem. The large sample capacity of the capillary cell, compared to the micro cell, produces sufficient desorption signal to often make the thermal diffusion effect negligibly small.

Lowell [144] has published a method to circumvent the problem of thermal diffusion by using an adsorbate with a low vapor pressure, such as krypton, at

Table 16.1 Data obtained using conventional cell (measured using the single-point BET method with 20% N_2 in He)

Weight	Actual area (m²)	Measured area (m²)	Deviation (%)	Signal shape at start of desorption peak
1.305	5.07	5.07	0	
0.739	2.87	2.87	0	
0.378	1.47	1.45	1.4	
0.177	0.678	0.686	−1.2	
0.089	0.345	0.327	5.2	
0.049	0.190	0.166	12.6	
0.0190	0.0730	0.0481	34.1	
0.0101	0.0394	0.0192	51.3	

liquid nitrogen temperature. The coefficient of thermal diffusion $D(t)$ is given by [145]

$$D(t) = \frac{N_1/N_t - N_1'/N_t}{\ln T_1/T_2} \qquad (16.12)$$

where N_1 and N_1' are the adsorbate concentrations at the absolute temperatures T_2 and T_1, respectively with $T_2 > T_1$. The term N_t is the total molecular concentration of adsorbate and carrier gas. Because of krypton's low vapor pressure, its mole fraction in the BET range of relative pressures is of the order of 10^{-4}. This small value causes the difference between N_1/N_t and N_1'/N_t nearly to vanish, with the consequence that no obscuring thermal diffusion signals are generated.

Attempts to increase the size of nitrogen adsorption or desorption signals, by using larger sample cells, results in enhanced thermal diffusion signals due to the increased void volume into which the helium can settle. However,

Table 16.2 Data obtained using U-tube cell

Wt ZnO (g)	Actual area (m²)	Measured area (m²)[†]
0.059	0.228	0.230
0.027	0.105	0.102
0.0045	0.0175	0.0161

[†] This value is corrected by 15 cm² for the cell wall area, as estimated from the cell dimensions. Desorption peaks from an empty U-tube cell gave areas of 12–17 cm².

when krypton is used, no thermal diffusion effect is detectable in any of the sample cells shown in Fig. 16.10.

The adsorption signals using krypton–helium mixtures are broad and shallow because the adsorption rate is limited by the low vapor pressure of krypton. The desorption signals are sharp and comparable to those obtained with nitrogen since the rate of desorption is governed by the rate of heat transfer into the powder bed.

With krypton, the ability to use larger samples of low area powders facilitates measuring low surface areas because larger signals are generated in the absence of thermal diffusion. Also, as is true for nitrogen, krypton measurements do not require void volume, or ideality corrections, nor is thermal transpiration a factor as in the volumetric measurements.

16.8 DATA REDUCTION–CONTINUOUS FLOW

Table 16.3 can be used as a work sheet for calculating specific surface area from continuous flow data. The data in the lower left corner is entered first and is used to calculate other entries. In the example shown, nitrogen is the adsorbate.

Column 1 is the mole fraction of adsorbate in the flow stream. Column 2 is obtained as the product of P_a and column 1. Column 2 when divided by column 3 gives the relative pressure, which is entered in column 4 and from which columns 5 and 6 are calculated.

Column 7 is the volume required to calibrate the desorption signal and column 8 is the corresponding weight of the calibration injection, calculated from the equation in the lower left side of the work sheet. The terms A_s and A_c are the areas under the signal and calibration peaks, respectively.

Columns 11–13 are calculated from the data in the previous columns. The data in column 13 is then plotted versus the corresponding relative pressures in column 5. The slope s and intercept i are calculated and the value of W_m is found as the reciprocal of their sum. Equation (4.13) is used to obtain the total sample surface area S_t and dividing by the sample weight yields the specific surface area, S.

16.9 SINGLE POINT METHOD

The assumption of a zero intercept reduces the BET equation to equation (5.3). This assumption is, of course, not realizable since it would require a BET C value of infinity. Nevertheless, many samples possess sufficiently high C values to make the error associated with the single-point method acceptably small (see Chapter 5 and Table 5.1).

Using the zero intercept assumption, the BET equation can be written as

$$W_m = W\left(1 - \frac{P}{P_0}\right) \qquad \text{(cf.5.3)}$$

from which the total surface area can be calculated by

$$S_t = W\left(1 - \frac{P}{P_0}\right)\frac{\bar{N}}{\bar{M}}\mathscr{A} \qquad \text{(cf.5.4)}$$

From the ideal gas equation of state

$$W = \frac{P_a V \bar{M}}{RT} \qquad (16.13)$$

so that

$$S_t = \left(1 - \frac{P}{P_0}\right)\frac{P_a V \bar{N}}{RT}\mathscr{A} \qquad (16.14)$$

where P_a and T are the ambient pressure and absolute temperature, respectively, $\bar{N}$ is Avogadro's number, $\mathscr{A}$ is the adsorbate cross-sectional area, and V is the volume adsorbed.

Using nitrogen as the adsorbate at a concentration of 0.3 mole fraction and assuming P_0 is 15 torrs above ambient pressure, equation (16.14) can be expressed as

$$S_t = \left(1 - \frac{0.3 P_a}{P_a + 15}\right)\frac{P_a V \bar{N}}{RT}\mathscr{A} \qquad (16.15)$$

Assuming that ambient pressure P_a is 760 torrs and ambient temperature T is 295 K, equation (16.15) reduces to

$$S_t = (2.84) V \, m^2 \qquad (16.16)$$

Thus, the total surface area contained in the sample cell is given by the simple linear relationship above when V is in cubic centimeters.

Table 16.3 Multipoint BET surface area data sheet

Date: _____   Sample: Titania   Total Weight: 12.0434 g

Operator: _____   Outgas Procedure: He purge, 1 hr, 150°C   Tare: 11.3430 g

Sample Weight (W): 0.7004 g

1	2	3	4	5	6	7	8	9	10	11	12	13
X_{N_2}	P (torr)	P_0 (torr)	P/P_0	P_0/P	$(P_0/P)-1$	V_c (cm³)	W_c (g)	A_s	A_c	$W = \dfrac{A_s W_c}{A_c}$	$W[(P_0/P)-1]$	$\dfrac{1}{W[(P_0/P)-1]}$
0.050	37.68	768.6	0.0490	20.398	19.398	2.00	0.00230	1066	1083	0.00226	0.0438	22.81
0.100	75.36	768.6	0.0980	10.199	9.199	2.20	0.00252	1078	1050	0.00259	0.0238	41.97
0.200	150.7	768.6	0.1961	5.096	4.096	2.50	0.00287	1168	1098	0.00305	0.0125	80.05
0.2995	225.7	768.6	0.2937	3.405	2.405	2.60	0.00298	1031	879	0.00350	0.0084	118.80

Ambient pressure, $P_a = 753.6$ mm Hg

Vapor pressure, $P_0 = 768.6$ mm Hg

Ambient temperature, $T = 295$ K

Adsorbate molecular weight, $\overline{M} = 28.01$

Adsorbate area, $\mathscr{A} = 16.2 \times 10^{-20}$ m²

Calibration gas weight, $W_c = \dfrac{P_a \overline{M} V_c}{(6.235 \times 10^4) T}$

Plot $\dfrac{1}{W[(P_0/P)-1]}$ vs P/P_0

Slope, $s = 391.9$

Intercept, $i = 3.52$

$W_m = \dfrac{1}{s+i}$

$W_m = 0.00253$ g

Total surface area,

$$S_t = \dfrac{W_m (6.023 \times 10^{23}) \mathscr{A}}{\overline{M}}$$

$S_t = 8.81$ m²

$S = S_t/W$

$S = 12.58$ m² g^{-1}

196 Dynamic methods

By calibrating the desorption signal A_d with a known volume of nitrogen V_c, equation (16.16) can be rewritten as

$$S_t = (2.84)\frac{A_d}{A_c}V_c \qquad (16.17)$$

where A_d and A_c are the integrated areas under the desorption and calibration signals, respectively.

A commercial single point instrument [146] contains a linearization network which corrects for the hot wire detector nonlinearity. This procedure allows a built-in digital integrator to integrate the signals linearly so that the surface area is given directly on a digital display. Analysis time on this apparatus is extremely short, usually under ten minutes.

17
Other flow methods

17.1 PRESSURE JUMP METHOD

A modification of the Nelson and Eggertsen method was made by Haley [147] who also used helium and nitrogen in a continuous flow system. Haley's apparatus differs from that described in the previous chapter in that the sample cell can be pressurized while the effluent flows from a pressure regulator through the cell to a flow controller and then to the detector maintained at ambient pressure.

Starting with a 10% nitrogen in helium mixture, the sample, held slightly above ambient pressure, is immersed in the coolant. When the adsorption signal is complete, the pressure is raised to two atm, which increases the partial pressure of nitrogen to 0.2 atm or the equivalent of a 20% mixture at 1.0 atm. After the completion of each adsorption signal, the pressure is incrementally increased until, at 1 MPa (150 PSIA) saturation is reached.

Desorption is accomplished by incrementally decreasing the pressure and monitoring the nitrogen-rich desorption signals.

Haley used a gas calibration loop of fixed volume to calibrate the signals. Under pressure as high as 1 MPa, septums and syringes are prone to leakage and injections of large volumes are difficult to achieve due to the opposing pressure. Also, the detector's nonlinearity requires that the calibration signal closely match the adsorption or desorption signal so that a fixed volume calibration loop is questionable in this application.

The entire apparatus must be very tight to prevent leakage at high pressures. The sample cell should be metal or heavy wall glass and is wrenched into position. A further difficulty is that at high pressures the linear flow velocity through the sample cell is small. For example, to maintain the same volumetric flow rate through the detector at 1 MPa will require one-tenth the linear flow rate as at 0.1 MPa because of gas expansion to atmospheric pressure. Therefore, the signal duration is excessively long, being broader and shorter as the pressure increases. Increasing the flow rate makes the signals narrower but their height remains small due to the dilution within the sample cell at high pressures.

17.2 CONTINUOUS ISOTHERMS

Semonian and Manes [148] have devised an approach which provides continuous data from which the desorption isotherm can be constructed. Their method utilizes a calibrated thermal conductivity detector for sensing the effluent concentration from a cell filled with adsorbate and slowly purged with a carrier gas. The amount desorbed at any relative pressure is calculated by integrating the effluent flow rate and thermal conductivity signal.

The tail end of the signal approaches the zero signal axis asymptotically, making accurate integration difficult and extending the time required for analysis. This difficulty is overcome by allowing the cell temperature to rise in order to hasten the desorption rate at the tail end of the curve. Using the Polanyi potential theory, Semonian and Manes reduced the nonisothermal data to points on the isotherm by equating the adsorption potential at various temperatures to the potential at the isotherm temperature, viz.,

$$RT \ln \frac{P_0}{P} = RT' \ln \frac{P'_0}{P'} \qquad \text{(cf.9.2)}$$

The partial pressure P' at temperature T' is calculated from P and T, the measured values.

The method gave good agreement with static data using n-butane as the adsorbate, nitrogen as the carrier gas and activated carbon as the adsorbent.

17.3 FRONTAL ANALYSIS

Early theories of elution chromatography developed by Wilson [149], Glueckauf [150–152], De Vault [153] and others [154–155] established the foundations for the development and propagation of solute bands through a packed column. Stock [156] succeeded in testing these theories for the elution of volatile adsorbed substances from a packed column by an inert gas. Stock measured the effluent concentration using a thermal conductivity detector and constructed the isotherms from the resulting 'chromatograms'. Gregg [157] outlined the method giving several examples and an explanation of the theory. The method of frontal analysis has been used and discussed by Cremer and Huber [158]. Malamud, Geisman and Lowell [159] have measured the isotherms of o-, m-, and p-xylene isomers on zinc oxide using frontal analysis.

The following mathematical description [157] of the development of a chromatographic elution profile assumes that the vapor is in equilibrium with the adsorbent, that diffusion is absent or of negligible rate compared to the flow rate, and that the adsorbent is uniformly packed in the column.

Consider that the flow through a packed column is suddenly switched from pure carrier gas to a mixture of carrier gas and adsorbate vapor at concentration c. As the mixture traverses the distance $d\ell$ along the column, adsorption will cause the concentration to decrease by dc. Expressing the concentration as weight W of adsorbate per unit volume of mixture, gives the weight lost by the flow in the column length $d\ell$ as

$$dW = \left(\frac{\partial c}{\partial \ell}\right)_v d\ell dv \qquad (17.1)$$

where dv is the volume of the mixture which has traversed the distance $d\ell$.
The weight gained by the adsorbent in the length $d\ell$ is

$$dW = \left(\frac{\partial W}{\partial v}\right)_\ell dv d\ell \qquad (17.2)$$

Then, from equations (17.1) and (17.2)

$$\left(\frac{\partial W}{\partial v}\right)_\ell = \left(\frac{\partial c}{\partial \ell}\right)_v \qquad (17.3)$$

In general, for any isotherm

$$W = f(c) \qquad (17.4)$$

and

$$dW = f'(c) dc \qquad (17.5)$$

Then

$$f'(c) = \left(\frac{\partial c}{\partial \ell}\right)_v \left(\frac{\partial v}{\partial c}\right)_\ell = \left(\frac{\partial v}{\partial \ell}\right)_c \qquad (17.6)$$

Taking the reciprocal of the preceding yields

$$\frac{1}{f'(c)} = \left(\frac{\partial \ell}{\partial v}\right)_c \qquad (17.7)$$

Equation (17.7) asserts that a volume of the mixture of adsorbate and carrier gas, at concentration c, moves through the column at a rate which is inversely proportional to the slope of the isotherm at that concentration.

Isotherms of Type I are characterized by a continuously decreasing slope which often becomes zero. Therefore, the adsorbate will move more rapidly through the column at higher concentrations. When the flow is switched from a mixture to pure carrier gas, the elution profile or chromatogram will appear as shown in Fig. 17.1, when plotted on a strip chart recorder.

In Fig. 17.1, P_0 is the signal produced by saturated vapor, t_1 is the time of switch-over to pure carrier gas. At time t_2 the signal corresponds to the pressure P and the horizontally shaded portion of the curve is proportional to the quantity of adsorbate remaining in the column. By successive integrations of the area under the curve along the line SPF, the quantity adsorbed at various relative pressures can be calculated.

Normalization of the signal is achieved by measuring the signal height produced by saturated vapor of known vapor pressure in the carrier gas. The area produced on the chart record for a constant flow rate over a known time interval at saturation generates all the information required for calibration. To reduce a chromatogram to the corresponding isotherm, the weight adsorbed per unit weight of adsorbent, at a particular vapor pressure, is related to the chromatogram signal area by equation (17.8) below

$$\frac{W}{M} = \frac{PF\bar{M}A}{SCTRP_a} \tag{17.8}$$

where
- W = grams adsorbed
- M = grams of adsorbent in the column
- $\bar{M}$ = relative molecular mass of adsorbate
- A = area under the signal curve at a given pressure (sq. inches)
- F = flow rate (cm^3 min^{-1})
- P = adsorbate vapor pressure (mm Hg)
- S = signal height when $P/P_0 = 1$ (inches)
- C = chart speed (inches min^{-1})

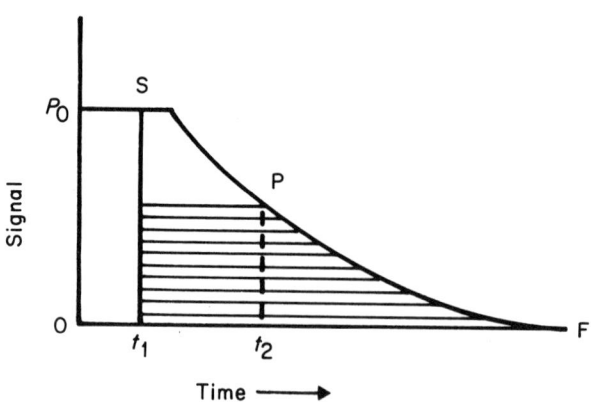

Figure 17.1 Elution profile for type I isotherms.

T = absolute temperature (K)
R = gas constant, 82.1 cm³ atm deg⁻¹ mol⁻¹
P_a = standard pressure, 760 mm Hg

The values of W/M calculated from equation (17.8) must be corrected for the weight of adsorbate contained in the column void volume, which can be obtained by subtracting the powder volume from the volume of the empty column.

Type III isotherms possess a continuously increasing slope. Therefore, the adsorbate propagates more rapidly at lower concentrations. The corresponding elution profile when the flow is switched from pure carrier gas to a mixture of carrier and adsorbate is shown in Fig. 17.2.

The times t_1, t_2 and t_3 in Fig. 17.2 correspond, respectively, to the time when the flow is switched from carrier to mixture, when the signal begins to emerge, and when the concentration of the adsorbate has risen to pressure P. The horizontally shaded area is proportional to the quantity remaining in the column at pressure P. By successively integrating the areas between the line $t_2\mathrm{PF}$ and the saturated pressure signal, the adsorbed quantity can be calculated at various relative pressures. Gregg [157] discusses the development of the Types II, IV and V isotherms which are more complex than the examples given here. However, it interesting to note that a Type II isotherm, for example, can be considered equivalent to a Type I at the lower end and a Type III at the upper end. Therefore, it is possible to obtain the entire isotherm in sections. By examining the rear of the elution profile after loading the column to a relative pressure of about 0.4, the lower part of the isotherm can be obtained while the upper section is determined after charging the column with saturated vapor and examining the front of the elution profile.

For isotherms that exhibit hysteresis, this method is of questionable value, since the high relative pressure end is measured as an adsorption isotherm but the low pressure end is measured as a desorption isotherm.

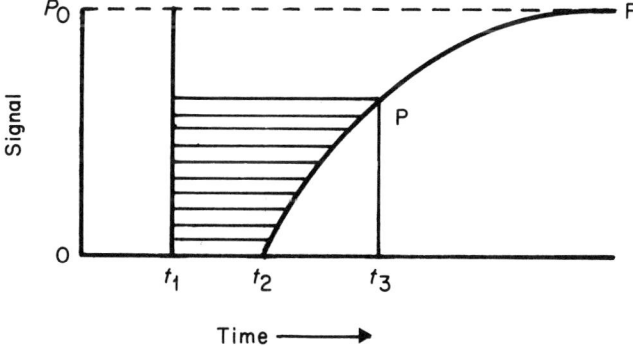

Figure 17.2 Elution profile for type III isotherms.

18
Gravimetric method

18.1 ELECTRONIC MICROBALANCES

Balances with adequate sensitivity to measure the small weight changes associated with adsorption can be divided into two categories — electronic beam microbalances and spring balances. Electronic beam balances usually use an electric current to restore the beam to the horizontal position as the weight changes. Their sensitivities are in the order of 1 µg with loadings of approximately one gram. Commercial electronic microbalances are available in vacuum containers with appropriate fittings for vacuum pumps and gauges.

Fig. 18.1 is a schematic of the Cahn [160] vacuum microbalance. The sensitivity is 1 µg with a load of 0.5 gram. Counterweight A is used to balance the adsorbent contained in bucket G which is immersed in a liquid nitrogen-filled Dewar flask H. Photocell C senses movement of the beam and provides current to the coil B to restore the beam. A metal plate D is fastened against the ground glass end of the vacuum bottle with electrical connections for bringing the signal to a recorder in order to monitor the weight changes continuously. Hang-down wire F is made of fine stainless steel. Ground joint E leads to pumps and gauges. For larger weights, the inner hook K can be used.

18.2 BUOYANCY CORRECTIONS

When using beam microbalances, the effects due to buoyancy and thermal transpiration must be considered. The gas displaced by the sample produces a buoyancy force that reduces the weight by exactly the mass of displaced gas. Buoyancy affects both the weighed mass and the counterweight. Also, the buoyancy changes with pressure. The mass of gas displaced is given by

$$W = \frac{\overline{M}PV}{RT} \tag{18.1}$$

Allowing V_w and T_w to be the volume and absolute temperature of the counterweight, and V_c and T_c to be the same for the sample and its container, the buoyancy correction becomes

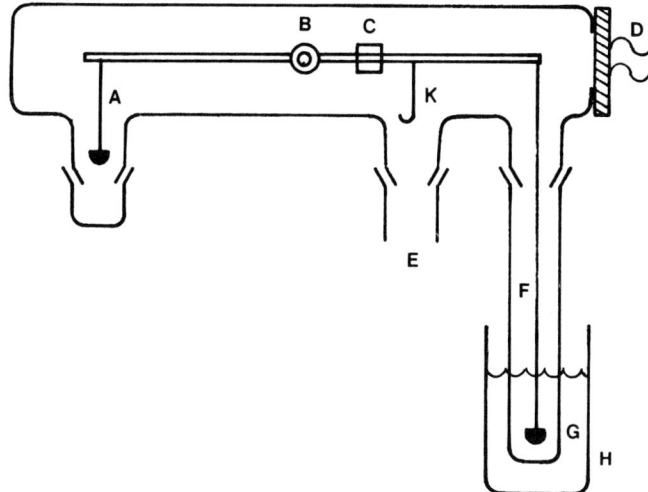

Figure 18.1 Cahn vacuum microbalance.

$$\Delta W = \frac{P\bar{M}}{R}\left(\frac{V_c}{T_c} - \frac{V_w}{T_w}\right) \qquad (18.2)$$

In order to estimate the error introduced by buoyancy with nitrogen as the adsorbate, allow T_w and T_c to equal 295 K and 77 K, respectively, and assume $V_w = V_c = V$, equation (18.2) then becomes

$$\Delta W = 0.00327\, PV \qquad (18.3)$$

with P in atmospheres and V in cm^3. In a completed monolayer, 1 µg of nitrogen will occupy 34.8 cm^2 of surface. Therefore, if the volumes V_c and V_w are each 2 cm^3, an error of 0.1 torr in pressure measurement will produce a weighing error of 0.86 µg or a surface area error of 29.9 cm^2. For 1% accuracy, it would then be necessary for the sample to exhibit an area not less than 2990 cm^2 (0.299 m^2).

The error caused by pressure measurement is not nearly as significant as that produced by measurements of the sample and counterweight volumes. By evacuating the container and admitting helium to a known pressure, say 300 torrs, the term in parenthesis of equation (18.2) can be evaluated. With this information, ΔW for nitrogen can be calculated at any desired pressure. If the buoyancy effect with helium can be evaluated to within 1%, then the error, ΔW for nitrogen, assuming V in equation (18.3) is 2.5 cm^3 and P is 200 torrs near the region of monolayer coverage, would be

$$\Delta W = 0.00327 \left(\frac{200}{760}\right)(2.5)(0.01) = 21.5\,\mu g \tag{18.4}$$

The corresponding surface area error would be $748\,cm^2$. Thus, to insure accuracy of 1% or better it would be necessary to measure no less than about $7.5\,m^2$.

18.3 THERMAL TRANSPIRATION

As shown previously (Section 15.7), thermal transpiration or thermomolecular flow arises when gas is contained in two vessels at equal pressures but different temperatures and a connection is made between the vessels. A vacuum microbalance represents just such a system. The long tube surrounding the hang-down wire separates the sample, immersed in the coolant bath, from the warm upper portion of the apparatus. With vacuum microbalances, the effects of thermal transpiration are further exacerbated by the temperature gradient along the hang-down wire.

When using low vapor pressure adsorbates, such as krypton at liquid nitrogen temperature, the entire isotherm lies within the thermal transpiration region. The effect of thermal transpiration is to produce an equilibrium uptake at a pressure less than that indicated on the manometer. The microbalance will indicate a weight change as thermomolecular flow occurs in the range of 0.001 to 20 torrs. The extent of the weight change depends on the physical dimensions of the tube enclosing the sample, the size of the sample, thermal conduction of the hang-down wire, the gas used, and the pressure.

With nitrogen, the data required for BET analysis or pore volumes lies above the thermal transpiration region.

When using microbalances for adsorption measurements, those adsorbates which do not require thermal transpiration corrections are the most susceptible to buoyancy errors while those adsorbates not requiring buoyancy corrections, such as krypton, because of its low vapor pressure, are most susceptible to thermal transpiration errors.

18.4 OTHER GRAVIMETRIC METHODS

Helical spring balances first used by McBain and Bakr [161] have been extensively used for adsorption measurements. The spring is suspended inside a glass tube by attachment to a hook at the top. A bucket containing the adsorbent is connected to the bottom of the spring. The bottom of the tube, containing the sample, is immersed in the coolant. The upper portion of the tube is connected to a vacuum pump, source of adsorbate, and manometer.

The spring is calibrated by adding known weights and measuring the extension with a cathetometer. The upper and lower portions of the tube should be thermostated to retain the spring calibration. Temperature coefficients of extension can amount to 10% of the spring's sensitivity, which is about 1 mg mm^{-1} with silica springs, which are commonly employed because of their linearity, high sensitivity, and reversibility.

Many improvements in design and material have been made since McBain and Bakr's original paper. Gregg and Sing [14], as well as Young and Crowell [162], review some of the more recent developments of spring as well as beam balances.

Other gravimetric techniques include flowing a mixture of carrier gas and adsorbent through a tube packed with adsorbent. Stopcocks at each end of the tube permit isolation of the sample. The tube is then removed from the flow system and weighed. Two successive equal weights indicate that equilibrium is achieved. The quantity adsorbed must be corrected for the gas contained in the void volume.

19
Comparison of experimental adsorption methods

The following comparisons of the three most common adsorption techniques (volumetric, continuous flow, and gravimetric) are based on the assumption that routine measurements are to be made. Special requirements may oblige the experimenter to choose one method in preference to another.

Ideal gas corrections

Volumetric measurements require that adsorbed volumes be corrected for ideality. The continuous flow method does not require this correction because the detector senses the gas concentration change at ambient temperature and pressure. Gravimetric apparatus measures the adsorbed weight directly and needs no correction for deviations from ideality.

Void volume measurements

The void volume must be measured in volumetric systems. The continuous flow and gravimetric methods are independent of the void volume.

Buoyancy corrections

Buoyancy corrections are only required by the gravimetric method.

Speed

Volumetric systems approach equilibrium asymptotically. The heat of adsorption is dissipated primarily through the powder bed at low pressures in the BET range. Continuous flow apparatus are more rapid because a fresh supply of adsorbate is constantly brought to the adsorbent, thereby driving the system to equilibrium. Heat is dissipated rapidly through the powder bed as well as through the flowing gas mixture which is at ambient pressure and contains helium. Gravimetric instruments approach equilibrium asymptotically as does the volumetric apparatus, but heat dissipation tends to be slow, particularly at low pressures where the principal source of thermal conduction is the hang-down wire.

Data points

The gravimetric and volumetric methods involve dosing the sample with adsorbate; the system subsequently comes to an equilibrium pressure which depends on the dosing volume, the isotherm shape, and the quantity of adsorbent. The continuous flow method produces data at the concentration of adsorbate in the flow stream. Therefore, the exact position of the data point can be chosen.

Thermal transpiration

Both the volumetric and gravimetric apparatus are subject to this effect. Data acquired at low pressures must be corrected for thermal transpiration. The continuous flow method is not subject to this phenomenon.

Thermal diffusion

The continuous flow method uses gas mixtures and is, therefore, the only one of the three methods subject to this effect. It occurs only when low areas are measured and can be eliminated by using an adsorbate with low vapor pressure such as krypton.

Liquid nitrogen level

Gravimetric and continuous flow equipment require only that the coolant level be sufficiently high as to maintain the sample at bath temperature. The volumetric method requires that the level be maintained constant in order to avoid changes in the void volume.

Thermostating

For maximum accuracy, the manifold and calibrated volumes in a volumetric apparatus should be maintained at constant temperature. Thermostating is not necessary for vacuum microbalances but in helical spring balances the spring should be maintained at constant temperature. Continuous flow apparatus need not be thermostated since the signals are immediately calibrated with know volumes at the same temperature and pressure. However, ambient temperature and pressure must be known to insure accurate calibration.

Gas mixtures

Only the continuous flow method requires mixed gases. Gas mixtures can be purchased commercially and can be prepared to accuracies of 1% relative. Alternatively, flows can be blended but often with some loss in accuracy.

Vacuum

The gravimetric and volumetric apparatus both require vacuum systems. The continuous flow method does not.

Degassing

Because the gravimetric and volumetric apparatus use vacuum systems it is convenient to degas by vacuum. Degassing in a continuous flow apparatus is accomplished by purging.

Calibration

Volumetric systems are calibrated by weighing the amount of mercury required to fill internal volumes. Gravimetric systems are calibrated using reliably known weights and should be recalibrated occasionally to insure that the sensitivity has not drifted with use. Continuous flow instruments are calibrated with gas sampling loops or syringes, the accuracy of which can be determined by filling with mercury and weighing.

Sample size

Gravimetric systems are limited to the maximum loading of the balance, usually about one gram. Volumetric apparatus should use samples sufficiently large to reduce the void volume error to an acceptable percentage. Usually a sample with about $10\,m^2$ of surface area is used with nitrogen. Flow systems can accommodate a wide range of sample size and are well suited to small quantities of sample.

Fragility

By their nature, gravimetric apparatus are quite fragile and care must be exercised in their use. Volumetric equipment usually contains fragile glass components. Continuous flow instruments are all metal, except for the sample cell, and are the least fragile of the various apparatus.

Pressure overshoot

The dosing requirement of gravimetric and volumetric apparatus can lead to pressure overshoot which may produce data off the isotherm in the hysteresis region. The continuous flow method is not susceptible to this phenomenon.

Adsorbates

Gravimetric instruments can accommodate any adsorbate. Continuous flow instruments are limited to adsorbates which will not condense at room temperature. Volumetric apparatus require various internal volume and manometer alterations when adsorbates with low vapor pressures are used.

Relative pressure

Saturated vapor pressures are determined independently of the adsorption method employed. The adsorbate pressure must be measured by a manometer or gauges when gravimetric or volumetric apparatus are used. Barometric pressure must be known for continuous flow measurements.

Sensitivity

The thermal conductivity detector used in the continuous flow method can sense signals corresponding to less than $0.001\,\mathrm{cm}^3$ of adsorption with 1% accuracy, causing it to be considerably more sensitive to small amounts of adsorption than the volumetric or gravimetric methods.

Permanent record

Vacuum electronic microbalances and continuous flow instruments provide electronic signals for a permanent record.

Errors

Both the gravimetric and volumetric methods are subject to cumulative errors. The continuous flow method produces each data point independently and is not subject to error accumulation.

20
Chemisorption

20.1 INTRODUCTION

The various forces responsible for physical adsorption were discussed in Chapter 2. These include dispersion as well as coulombic forces. Chemical adsorption arises from the inability of surface atoms to interact symmetrically in the absence of a neighbor above the plane of the surface. For this reason, surface atoms often possess electrons or electron pairs which are available for bond formation.

Because chemisorption involves a chemical bond between the adsorbate and adsorbent, only a single layer of chemisorption can occur. Physical adsorption on top of the chemisorbed layer and diffusion of the chemisorbed layer beneath the surface can obscure the fact that chemisorbed material can be only one layer in depth.

Often an attempt is made to distinguish between physical and chemical adsorption on the basis of the heat of adsorption. However, this is not an entirely satisfactory procedure. The smallest physical heat of adsorption will be slightly greater than the heat of liquefaction of the adsorbate. Were this not true, the vapor would condense and not be adsorbed. The upper limit for physical adsorption may be higher than 20 kcal mol^{-1} for adsorption on adsorbents with very narrow pores, such as silica gels and zeolites. The heats of chemisorption range from over 100 to less than 20 kcal mol^{-1}. Therefore, only very high or very low heats of adsorption can be used as criteria for the type of adsorption process. A more definitive criterion as to whether a particular interaction is physical or chemical is to search for reaction products. Other techniques, such as isotope exchange, electron diffraction, absorption spectroscopy, magnetic susceptibility, and electron spin resonance are just a few of the modern methods which can be used to establish a detailed description of the adsorbate–adsorbent interaction.

20.2 CHEMISORPTION EQUILIBRIUM AND KINETICS

The measurement of the equilibrium between the gaseous and the chemisorbed state is frequently difficult because of the very low equilibrium pressures required to saturate the surface. Often, in the case of strong interactions the monolayer is completed at very low pressures, even less than one torr, as shown in Fig. 20.1.

When an adsorbate molecule is strongly bound and localized to one adsorption site, as in chemisorption, the Langmuir equation is applicable

$$\frac{W}{W_m} = \frac{KP}{1+KP} \qquad \text{(cf.4.11)}$$

At sufficiently low pressure $KP \ll 1$, the Langmuir equation becomes

$$\frac{W}{W_m} = KP \qquad (20.1)$$

This region of the isotherm is the Henry's law region; the uptake is directly proportional to the pressure. At high pressures, $KP \gg 1$, so that Langmuir's equation becomes $W = W_m$.

Fig. 20.2 illustrates some of the essential differences between physical and chemical adsorption [163].

Curves I and II represent potential energy plots for chemical and physical adsorption, respectively. The zero of potential energy is taken, as is customary, at infinite separation of the interacting species. The minimum in curve I, below zero potential energy, is equal to the heat of chemisorption, ΔH_C and the minimum of curve II is equal to the heat of physical adsorption, ΔH_P. The fact that curve I lies above zero potential energy at large internuclear separations implies that the chemisorbed gas is in an activated state or has undergone dissociation. The term ΔH_D, then, is the heat of activation or dissociation. If activation or dissociation does not occur, then curve I would approach zero potential energy asymptotically, similar to curve II. The minimum of

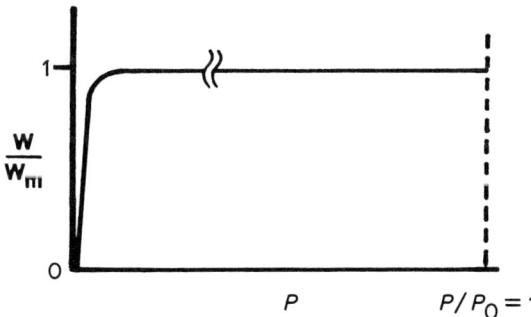

Figure 20.1 Type I chemisorption isotherm for a strongly bound adsorbate.

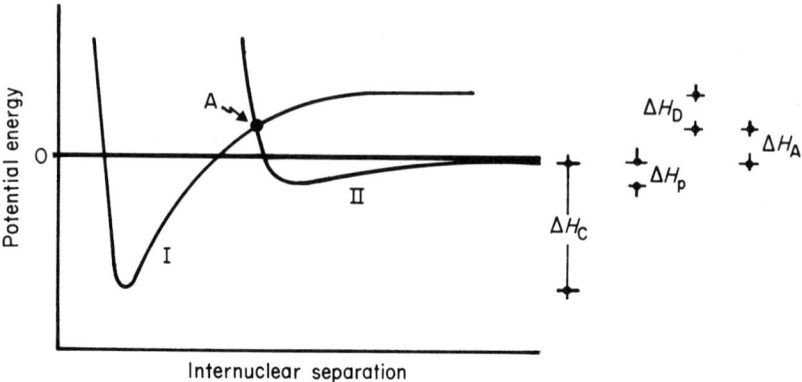

Figure 20.2 Potential energy curves for physical (II) and chemisorption (I).

curve I occurs at a smaller internuclear separation than that of curve II because chemical bonding, involving orbital overlap, will bring nuclei closer together than can the less energetic physical adsorption forces.

The transition from physical adsorption to chemisorption occurs at point A. The potential energy at A is in excess of that for the adsorbate and the adsorbent when separated and represents the activation energy required for chemisorption, ΔH_A. If curve I resided more to the right or curve II more to the left, then the transition from physical to chemical adsorption would occur with no activation energy since the crossover point would reside beneath zero potential energy.

The concept of activated adsorption explains why the heat of adsorption is often small at low temperatures and large at higher temperatures. This is because at low temperature the heat is due to physical adsorption whereas at higher temperature sufficient thermal energy is available to provide the required activation for chemisorption with its associated high heat of adsorption. Also, the rates of adsorption frequently decrease at higher temperatures indicating that the adsorption mechanism undergoes a transition to an activated process. A further frequent observation is that the amount adsorbed varies with temperature as shown in Fig. 20.3 [163]. The initial decrease is due to thermal desorption of the physically adsorbed gas. Subsequently, the quantity adsorbed increases with increasing temperature due to commencement of activated chemisorption. Finally, the curve slopes downward when sufficiently high temperature is reached to desorb the chemisorbed state thermally.

20.3 CHEMISORPTION ISOTHERMS

When the initial slope of a chemisorption isotherm is not large so that monolayer coverage is approached at conveniently measurable pressures, the

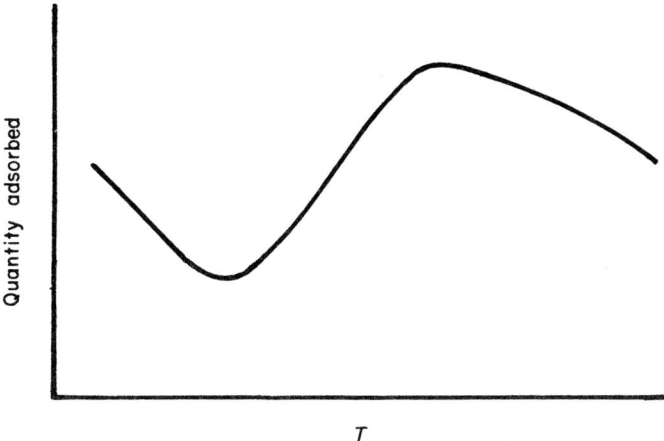

Figure 20.3 Variation in quantity adsorbed with temperature.

gravimetric, volumetric or continuous flow methods can be employed. The gravimetric and volumetric methods for determining this type of isotherm are carried out in the usual manner, provided that no reaction products contribute to the equilibrium pressure. The only essential difference when measuring chemical and physical adsorption using these techniques, is that in chemisorption the sample is maintained at an elevated temperature rather than in a coolant bath.

In the volumetric method, the chemisorption isotherm (Fig. 20.1) is acquired by measuring the gas uptake as a function of increasing pressure. The total chemisorbed volume is obtained by extrapolation of the linear portion of the isotherm to the Y axis. Assuming some stoichiometric relationship between the active adsorbent and the chemisorbed gas, the amount of dispersion of an active substrate on a support can be calculated.

When using the continuous flow method, however, some additional versatility is available in chemisorption measurements. For example, when data is required at an adsorbate pressure of 0.1 atm, a 10% mixture of adsorbate, mixed with an inert carrier gas, is passed through the apparatus with the sample cooled to a temperature at which no chemisorption can occur. Upon warming the sample to the required temperature, adsorption occurs producing an adsorbate-deficient peak that is calibrated by injecting carrier gas into the flow stream. Equation (16.9) is then used to calculate the quantity adsorbed. This process is repeated for each concentration required. Caution must be exercised to avoid physical adsorption when the sample is cooled to prevent chemisorption. Should this occur, the adsorption peak due to chemisorption can be obscured by the desorption peak of physically bound adsorbate when the sample is heated.

When there is no satisfactory temperature at which neither physical nor chemical adsorption will occur, an alternate method can be used. Referring to Fig. 16.9, valve V_6 can be rotated to shunt the gas flow past the cell. After flowing pure carrier gas through the sample cell and purging out any contaminants, the cell is isolated and the flow is switched, for example, to a 10% mixture. When a steady base line is obtained, valve V_6 is rotated to admit the mixture into the sample cell. Presuming that the sample is maintained at a temperature at which chemisorption will occur, a peak will be produced with an area proportional to the amount of chemisorption and to the volume of carrier gas swept out of the cell void volume. The cell is again isolated and the flow concentration increased, for example, to 20%. Again, when the flow is admitted into the sample cell, the area under the resulting peak will be proportional to the amount adsorbed plus the volume of 10% mixture swept out of the void volume. This process is continued until all the required data is obtained.

Because each signal must be corrected for the volume of the previous gas concentration contained in the cell, it becomes necessary to measure the cell void volume. This is accomplished by filling the cell containing the sample with an inert gas, then isolating the cell and flowing a second nonreacting gas through the system. When valve V_6 is rotated, the cell will be purged and the resulting peak is calibrated with a known volume of inert gas. The cell void volume V_v is given by

$$V_v = \frac{A}{A_{cal}} V_{cal} \tag{20.2}$$

where A is the area produced when the cell is opened, thereby purging the trapped gas, and A_{cal} is the area produced by injecting a known volume V_{cal} of the same gas. The volume adsorbed is then expressed as

$$V_{ads} = V_{carrier} \left(\frac{C_{adsorbate}}{C_{carrier}}\right)\left(\frac{A_{sig}}{A_{cal}}\right) - V_v \tag{20.3}$$

The terms in equation (20.3) are defined as follows: V_{ads} is the volume chemisorbed, measured at room temperature and pressure; $V_{carrier}$ is the volume of carrier gas injected when calibrating the peak produced by opening and purging the sample cell; $C_{adsorbate}$ and $C_{carrier}$ are the adsorbate and carrier gas concentrations in the flow stream. These concentrations are most conveniently expressed as mole fractions; A_{sig} is the area under the peak resulting from adsorption when the cell is opened and purged; A_{cal} is the area under the calibration curve using pure adsorbate; V_v is obtained from equation (20.2).

If two inert gases are not available, an empty cell can be calibrated using pure adsorbate and carrier gas. The void volume can then be calculated as the difference between the empty cell volume and the sample volume, presuming that the density of the sample is accurately known.

For maximum accuracy, those portions of the void volume not heated should be maintained at constant temperature to prevent the void volume from varying during the analysis.

Often chemisorption produces reaction products which are swept out of the sample cell to the detector. In such cases, to prevent obscuring the adsorption signal, it is necessary to remove the reaction products from the flow stream. Depending upon their nature they can often be removed chemically or by condensation in a cold trap.

When hydrogen is used as the adsorbate, helium will not adequately serve as a carrier gas because these two gases exhibit similar thermal conductivities. In this case nitrogen or any other nonreacting gas can serve as the carrier.

20.4 SURFACE TITRATIONS

When the adsorbate is strongly bound to the surface such that desorption does not occur or occurs at a negligibly slow rate, the monolayer capacity of the adsorbent can be obtained by titrating the surface with the adsorbate [164, 165]. This method requires flowing an inert gas over the sample and then injecting a quantity of adsorbate into the flow stream. As the band of adsorbate is carried over the sample, some adsorption will occur. The detector will respond to the quantity of adsorbate which is not chemisorbed. Subsequent adsorbate injections will produce successively larger peaks as the surface approaches saturation. The surface is saturated when two or more successive peaks exhibit the same signal area. The total volume adsorbed, V_{ads}, is given by

$$V_{ads} = V_1^{inj} - \frac{A_1}{A_1^{sat}} V_1^{inj} + V_2^{inj} - \frac{A_2}{A_2^{sat}} V_2^{inj} + \ldots + V_i^{inj} - \frac{A_i}{A_i^{sat}} V_i^{inj}$$

(20.4)

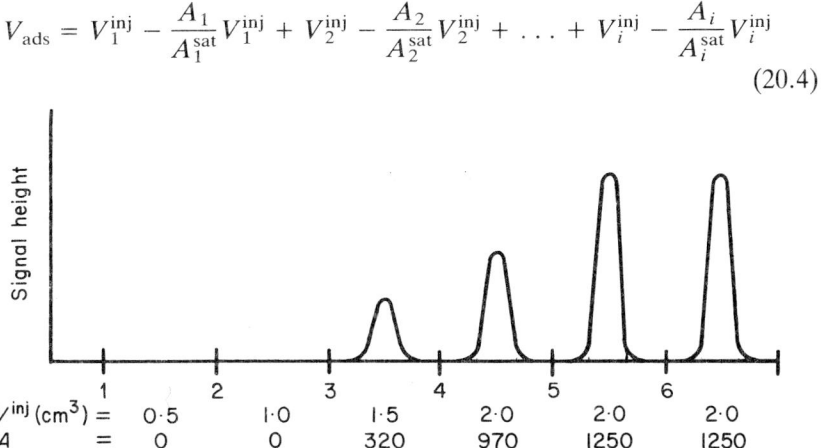

Figure 20.4 Chemisorption titration for the measurement of total uptake.

where V^{inj} is the volume of adsorbate injected into the flow stream, A is the signal area after the band of adsorbate is carried over the adsorbent, and A^{sat} is the signal area corresponding to the injected volumes when no adsorption occurs. Fig. 20.4 illustrates the quantity of uptake for a hypothetical case. For maximum accuracy a calibration curve should be prepared to eliminate the detector nonlinearity as a source of error.

21
Mercury porosimetry

21.1 INTRODUCTION

Although many high-pressure mercury porosimeters have been constructed, they all have several essential components, which are perhaps different in their design but nevertheless are common to each apparatus. These components include the following:

(a) Cell to hold the test sample.
(b) Dilatometer or stem attached to the sample cell in which the mercury level varies with intrusion or extrusion.
(c) Vacuum filling apparatus to remove trapped air from the pores and for transferring mercury into the sample cell.
(d) Pressure generator.
(e) High pressure vessel to contain the sample cell.
(f) Probe to measure the mercury level.
(g) Hydraulic fluid to transmit the pressure to the dilatometer.

21.2 PRESSURE GENERATORS

Commercial mercury porosimeters are available which achieve maximum pressure of 414 MPa corresponding to pore radii of 0.0018 µm, assuming 140° for the contact angle (see equation (11.1)). Although higher pressures can be achieved, the interpretation of data much above 414 MPa becomes questionable in view of the uncertainty regarding the meaning of contact angles and surface tension in pores not much wider than the diameter of a few mercury atoms.

Usually the required pressures are obtained with a reciprocating pressure intensifier which transmits a force over a wide area to a smaller area, thus increasing the applied pressure. Check valves are employed to direct the fluid flow into and out of the intensifier on each cycle. Because of the reciprocating operation of the pressure intensifier, most porosimeters operate on an incremental basis, that is, the intruded volumes are measured as discrete data points after each incremental change of pressure. Recently a new type of porosimeter has been developed employing a ram pressure generator which produces a continuous monotonic pressure ramp. This device will be discussed in Section 21.4.

21.3 DILATOMETER

A dilatometer is used to measure the intrusion and extrusion volumes and several methods are used to measure the change in mercury level within the dilatometer stem as intrusion and extrusion take place. One method involves the use of platinum or platinum alloy resistance wire placed coaxially in the stem of dilatometer tube. As the mercury level decreases, the amount of resistance wire exposed increases, thereby providing a voltage which increases linearly with decreasing mercury level. Very small current is used to minimize resistance heating of the mercury. Platinum or platinum–iridium alloy may be employed as the resistance wire because they do not amalgamate with mercury.

A second method for monitoring the change in mercury level is the use of a follower probe. As the level decreases, electrical contact between the probe and the mercury level is broken. This actuates a drive motor which moves the probe, through a pressure seal, until contact is reestablished. The drive motor angular displacement is proportional to the change in mercury level.

Fig. 21.1 illustrates a third method of measuring the mercury level. The mercury in the dilatometer stem constitutes one plate of a capacitor and a metal sheath coaxially surrounding the dilatometer serves as the second plate. The capacitance changes with the mercury level due to the variation of the effective plate area.

21.4 CONTINUOUS SCAN POROSIMETRY

A mercury porosimeter capable of producing continuous plots of both the intrusion and extrusion curves has recently been developed [146]. Fig. 21.2 is a schematic diagram of the continuous scan porosimeter shown in Fig. 21.3. A pressure generator, driven by a 1 hp motor through a 50 to 1 gear reduction, uses a threaded shaft to drive a stainless steel rod into a narrow cavity filled with hydraulic oil. As the rod penetrates into the cavity, the oil is compressed in the entire apparatus. The total volume displaced by the rod is 20 cm^3 which is sufficient to produce pressure up to 414 MPa on the mercury column. The rate of pressurization can be controlled by the motor speed. A microcomputer motor speed control is an integral part of the porosimeter design. It serves to adjust the pressurization or depressurization rate in inverse proportion to the rate of intrusion or extrusion. Thus, the porosimeter provides maximum speed in the absence of intrusion or extrusion and maximum resolution when most required, that is, when intrusion or extrusion is occurring rapidly with changing pressure. A strain gauge transducer is

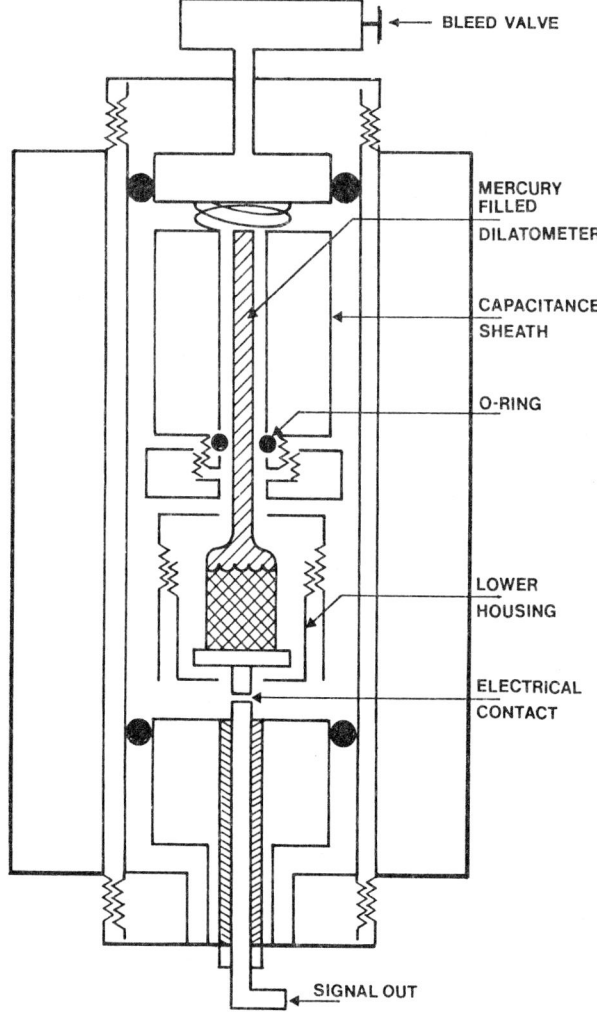

Figure 21.1 Capacitance probe; high-pressure cavity and seals for continuous intrusion–extrusion measurements.

located near the top of the generator and provides an electrical signal in proportion to the pressure.

The pressure vessel containing the sample cell is connected to the pressure generator with high pressure tubing. An oil reservoir near the top of the apparatus feeds oil to the pressure vessel as required for filling. The pressure vessel has removable caps on the top and bottom for loading the sample cell and for easy draining if a glass sample cell should break under test. The caps

220 Mercury porosimetry

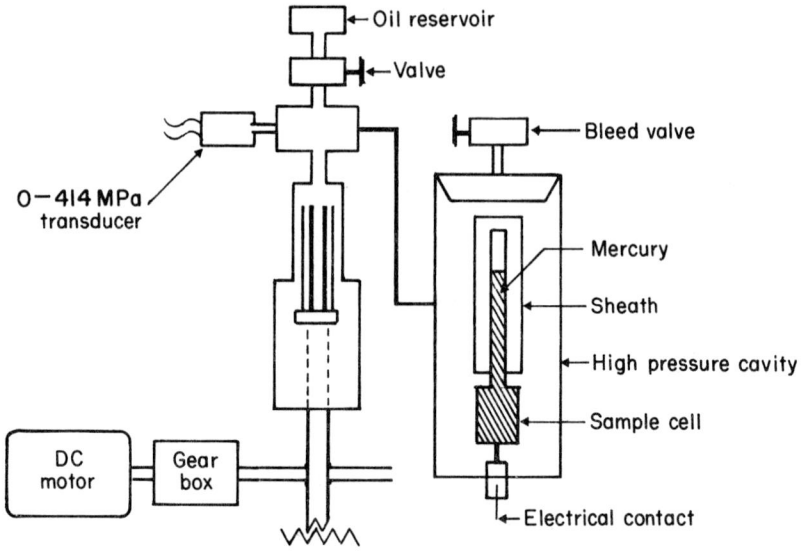

Figure 21.2 Continuous scanning mercury porosimeter.

are screwed into the pressure vessel and need be only finger-tight to retain 414 MPa.

The mercury level is monitored with a capacitance probe as shown in Fig. 21.1. The signal from the capacitance probe and from the pressure transducer are transmitted to a computer or plotted on an X-Y recorder as intrusion volume vs pore radius.

Some of the advantages of scanning porosimetry include the following:

(a) Continuous plots of both intrusion and extrusion curves are obtained.
(b) At nominal drive speeds, a complete analysis takes about 15 minutes.
(c) At no time along the intrusion curve does the pressure relax. Therefore, no opportunity is provided for a small quantity of extrusion to occur following intrusion. This is a condition which can produce a data point somewhere in the hysteresis region between the intrusion and extrusion curves.
(d) Unlike conventional incremental porosimeters, which produce a limited number of data points, the presentation of continuous scans eliminates the need to rerun an analysis to obtain information between data points for additional resolution in the regions of interest.
(e) The effects of mercury compression and the compressive heating of the hydraulic oil are thermodynamically compensated. Therefore, the need to make blank runs is unnecessary for all but the most exacting analysis. Blank runs made on cells filled with mercury show less than 1% of full scale signal over the entire operating range from 0 to 414 MPa.

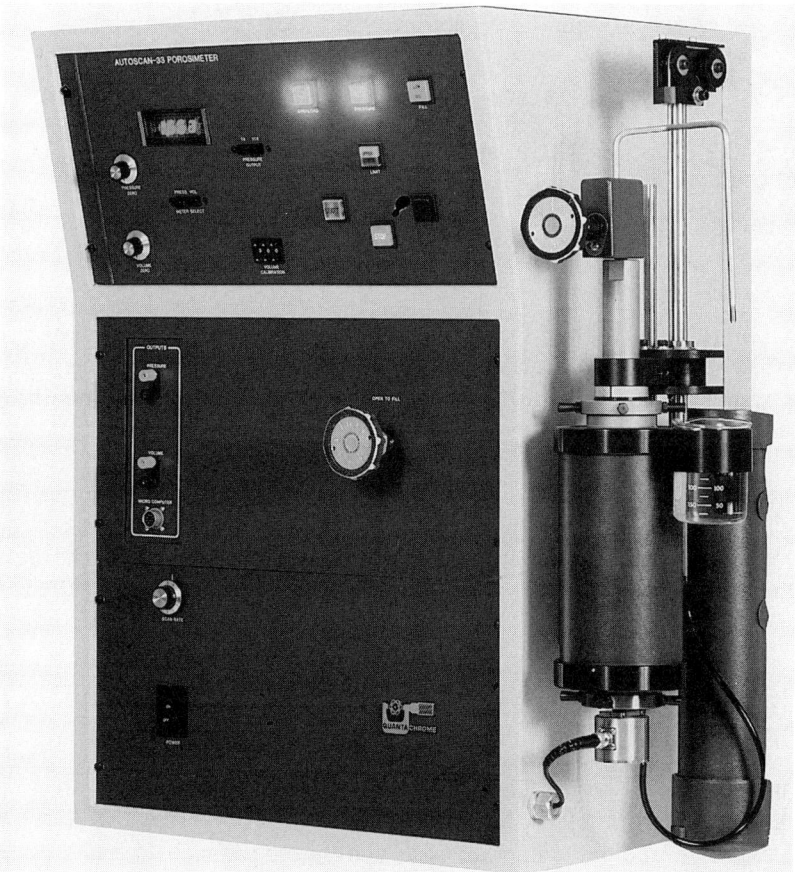

Figure 21.3 Autoscan-60 Porosimeter. Courtesy of Quantachrome Corporation.

21.5 LOGARITHMIC SIGNALS FROM CONTINUOUS SCAN POROSIMETRY

The signals produced by the strain gauge can be amplified linearly or logarithmically. In the linear mode the displacement of the X-Y recorder along the X or pressure axis is inversely proportional to the pore radius as given by equation (10.24)

$$Pr = -2\gamma\cos\theta \tag{cf.10.24}$$

When the strain gauge signal is fed to the recorder as the logarithm of the pressure the following relationships exist using A as $-2\gamma\cos\theta$

$$\log P = \log A - \log r \tag{21.1}$$

Since the electrical signal E is proportional to $\log P$ one can write

$$\log P = kE \tag{21.2}$$

Combining equations (21.1) and (21.2) and taking derivatives of each side, yield

$$kdE = -\frac{dr}{r} \tag{21.3}$$

Equation (21.3) is in the form of a first-order decay equation which describes a first order chemical reaction or radioactive decay. It establishes that the pore radius decreases logarithmically with increasing signal as described by equation (21.4), obtained by integrating equation (21.3)

$$r = Ke^{-kE} \tag{21.4}$$

Therefore, when semilog paper is used upside down in an X-Y recorder, so that the recorder pen sweeps from larger to smaller pore radii, the pore radius can be plotted directly.

21.6 LOW-PRESSURE INTRUSION–EXTRUSION SCANS

To measure pores larger than 7 μm (7 μm pores correspond to a pressure of about 1 atm) it is necessary to monitor intrusion at pressures less than 1 atm. Fig. 21.4 is a schematic diagram of a filling mechanism that can be used for monitoring intrusion up to 0.17 MPa, as well as for degassing the sample. The filling mechanism has a provision for heating the sample to hasten the rate of degassing.

After pumping to sufficiently low vacuum, as indicated by the thermocouple gauge, usually about 50 millitorrs, the glass vacuum housing is rotated toward the vertical. Mercury flows into the narrow end of the housing to cover the end of the dilatometer stem. The selector valve is rotated to admit compressed air or inert gas and the bleed rate is adjusted with the needle valve. When the sample cell and stem are filled with mercury, the bleed is stopped, the housing is returned to the horizontal position, and the pressure is read on the gauge. Thereafter, a small continuous bleed creates a pressure increase which forces mercury to intrude into the voids and large pores.

The dilatometer stem is coaxially enclosed in an open-ended stainless steel sheath which serves as one plate of a capacitor in a manner similar to the sheath used in the high-pressure porosimeter. When connected to the capacitance bridge of the porosimeter, the filling apparatus, using its own low-pressure transducer, measures the intruded volume and continuously plots the data up to 0.17 MPa.

Contact angle for mercury porosimetry

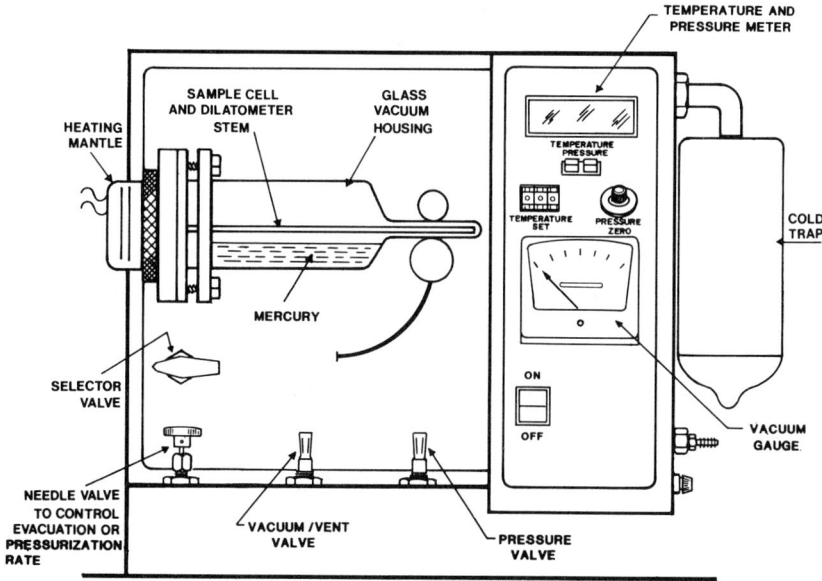

Figure 21.4 Filling mechanism for sample conditioning and low-pressure porosimetry.

Extrusion from the voids and large pores can be continuously monitored by evacuating the housing, surrounding the dilatometer stem, through a controlled bleed. This completes a full low-pressure intrusion–extrusion cycle. When the sample cell is removed from the filling mechanism it is inserted into the high-pressure cavity for high-pressure analysis. This requires that the cell be moved from the horizontal to the vertical position, which creates an additional pressure head of 0.03–0.04 MPa above ambient on the sample. Since the filling apparatus has acquired data up to 0.17 MPa, this additional pressure head proves to be no inconvenience.

21.7 CONTACT ANGLE FOR MERCURY POROSIMETRY

The accuracy of the measurement of pore radii, into which intrusion occurs, is limited by the accuracy to which the contact angle is known. For comparative purposes it is adequate to choose a reasonable value for the contact angle (130–140°) in order to ascertain whether or not two samples have the same pore size distribution and pore volume. However, if absolute data are required, it is necessary to measure the contact angle very accurately. While the great majority of materials exhibit contact angles near 140°, one must recognize that the nature of the cosine function is such that, at angles near 140°, the value of the cosine changes substantially with the angle. For example, an error of even 1° at 140° would introduce an error of slightly more

224 *Mercury porosimetry*

than 1.4% in $\cos\theta$ and, thus in the pore radius. Not only is it necessary to measure the contact angle accurately, but it should be measured under the same conditions that prevail in mercury porosimetry. This requires that the contact angle be measured as the advancing angle [166, 167] on a clean surface under vacuum and using conditions that match those of the actual intrusion as closely as possible.

A mercury contact anglometer that measures the contact angle on powders, meets the above requirements and gives highly reproducible results is shown in Fig. 21.5. A cylindrical hole of known radius is made in a compacted bed of powder. A press is used to compact the powder around a precision bore pin, which upon removal, produces a cylindrical open-ended pore.

Fig. 21.6 is a schematic of the cavity of the anglometer in which the contact angle is measured. As shown in the figure, B is the compressed bed of powder in the sample holder; A is the fixed volume of mercury placed above the powder to a known height and therefore, known pressure head; D is an O-ring seal to prevent the flow of gas from the upper to the lower chamber; E and F are evacuation ports for the two chambers; G is a piezoelectric crystal and H is a viewing lens.

After placing the housing with powder sample and mercury into the anglometer, both chambers are evacuated. Air is then allowed to bleed slowly into the upper chamber and the pressure is monitored by a sensitive transducer (not shown). A digital display, with an appropriate offset to allow for the mercury pressure head, indicates the pressure to the nearest 0.001 PSIA or

Figure 21.5 Mercury Contact Anglometer. Courtesy of Quantachrome Corporation.

Contact angle for mercury porosimetry 225

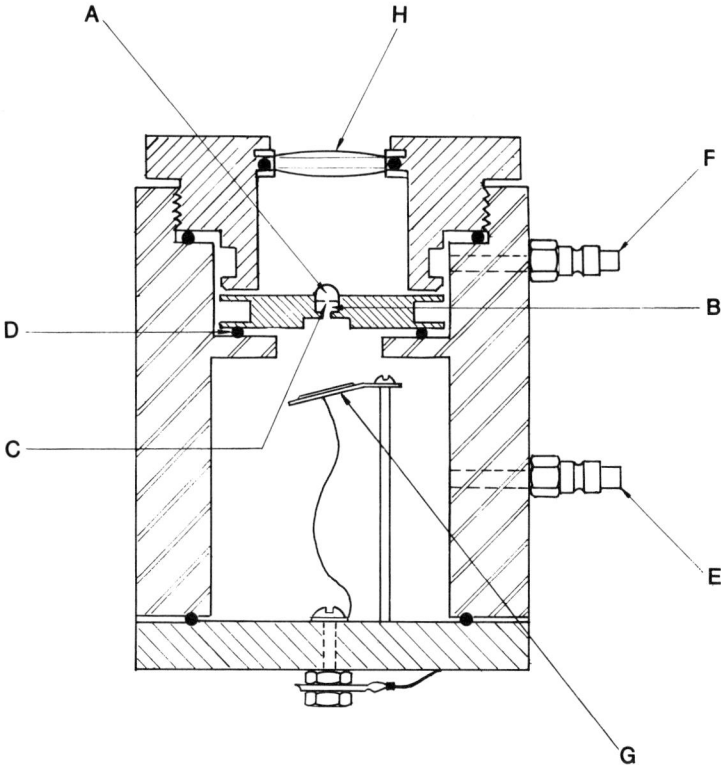

Figure 21.6 Anglometer cell housing.

the corresponding value of $\cos\theta$, which is proportional to the pressure. When the breakthrough pressure is reached, mercury is forced through the precision bore hole onto the piezoelectric crystal, which sends a signal to the digital display, locking its reading.

Table 21.1 illustrates the results of six successive contact angle measurements on each of ten powders, along with the reproducibility, expressed as standard deviation using the anglometer described above.

Using low-pressure porosimetry, Winslow [168] measured contact angles by determining the breakthrough pressure required to force mercury into numerous holes drilled into the surface of solid discs.

Another method [169] for measuring the contact angle utilizes a small drop of mercury placed on the surface of a smooth bed of powder which assumes a spherical shape except for the portion in contact with the surface. As additional mercury is added to the drop, the height increases until it reaches a maximum value. Further additions of mercury increase the drop diameter with no additional increase in the height. Fig. 21.7 illustrates the change of drop shape as mercury is added.

Table 21.1 Contact angle measurements of mercury on various materials

Material	Mean contact angle ($\theta°$)	Standard deviation
Dimethylglyoxime	139.6	0.45
Galactose	140.3	0.43
Barium chromate	140.6	0.41
Titanium oxide	140.9	0.55
Zinc oxide	141.4	0.34
Dodecyl sodium sulfate	141.5	0.44
Antimony oxide	141.6	0.88
Fumaric acid	143.1	0.27
Starch	147.2	0.68
Carbon	154.9	1.2

Figure 21.7 Change of mercury drop shape with size.

The relationship between contact angle θ and the maximum drop height h_{max} is

$$\cos\theta = 1 - \frac{\rho g h_{max}^2}{2\gamma} \qquad (21.5)$$

where ρ and γ are the density and surface tension of mercury, respectively, and g is the acceleration of gravity.

The maximum height method measures an equilibrium contact angle and not the advancing angle. The measurement is difficult to perform under vacuum conditions and further difficulties arise because the powder tends to float on the mercury drop while the drop is prone to sinking into the powder bed, unless it is tightly compacted.

Optical devices are sometimes employed for the measurement of contact angle, wherein the operator must attempt to establish the tangent to the contact angle of a drop of mercury resting on a plane surface. This method has never proven sufficiently accurate because of its inherent subjectivity. Different experimenters will inevitably measure substantially different contact angles and even the same person will observe different angles on the same material on different occasions.

22
Density measurement

In many areas of powder technology, the need to measure the powder volume or density often arises. For example, powder bed porosities in permeametry, volume specific surface area, sample cell void volumes as well as numerous other calculated values all require accurately measured powder densities or specific volumes. It is appropriate, therefore, to introduce some discussion of powder density measurements.

22.1 TRUE DENSITY

The true density is defined as the ratio of the mass to the volume occupied by that mass. Therefore, contribution to the volume made by pores or internal voids must be neglected when measuring the true density. Regardless of whether pores and internal voids are present, the density of fine powders is often not the same as that of larger pieces of the same material because in the process of preparing many powders, those atoms or molecules located near the surface are often forced out of their equilibrium position within the solid structure. On large pieces of material, the percentage of atoms near the surface is negligibly small. However, as the particle size decreases, this percentage increases with its resultant effect upon the density.

If the powder has no porosity, the true density can be measured by displacement of any fluid in which the solid remains inert. The accuracy of the method is limited by the accuracy with which the fluid volume can be determined. Usually, however, the solid particles contain pores, cracks or crevices which will not be completely penetrated by a displaced liquid. In these instances, the true density can be measured by using a gas as the displaced fluid. There are many commercially available instruments for the determination of the true density of powders, most of which operate on the principle of gas displacement. Helium is the most frequently used gas because the inertness and small size of the helium atom enables it to penetrate even the smallest pores.

Apparatus used to measure solid volumes are often referred to as pyknometers or pycnometers after the Greek 'pyknos' meaning thickness or

228 Density measurement

density. Once the sample volume and mass have been determined, the density is readily calculated.

The recent availability of economical, stable and high-resolution pressure transducers has led to the development of a new type of helium pycnometer which offers extreme simplicity of operation along with great speed and accuracy. A schematic diagram of this pycnometer is shown in Fig. 22.1. The volume V_c in the shaded area is considered the sample cell. After purging the system with helium and bringing all volumes to ambient pressure, the sample cell is pressurized to P_2, about 1 atm above ambient. Valve V_1 is opened to connect the reference volume V_R to that of the cell. Consequently, the pressure drops to a lower value P_1 in the cell while increasing from ambient to P_1 in the reference volume. Using P_1 and P_2, the powder volume V_p is calculated from

$$V_p = V_c + \frac{V_R}{1 - P_2/P_1} \qquad (22.1)$$

Equation (22.1) is derived as follows: At ambient pressure, P_a, the state of the system is described as

$$P_a V_c = nRT_a \qquad (22.2)$$

where n is the number of moles of gas occupying the volume V_c at P_a, R is the gas constant and T_a is ambient temperature. When a sample of volume V_p is placed in the sample cell, equation (22.2) can be rewritten as

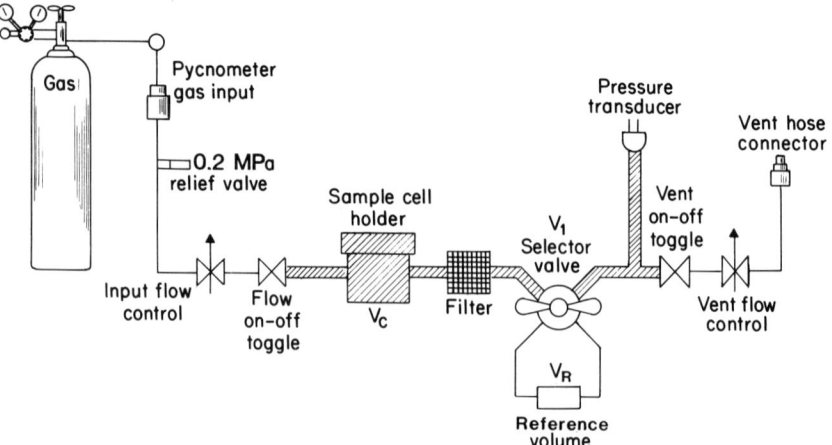

Figure 22.1 Pycnometer for true powder densities. Courtesy of Quantachrome Corporation.

$$P_a(V_c - V_p) = n_1 R T_a \tag{22.3}$$

where n_1 is the moles of gas in the sample cell at P_a and T_a. When the system is pressurized to P_2, it is described by

$$P_2(V_c - V_p) = n_2 R T_a \tag{22.4}$$

where n_2 is the moles of gas in the sample cell at the new pressure, P_2. After volume V_R is added to that of the sample cell and the pressure falls to P_1, then

$$P_1(V_c - V_p + V_R) = n_2 R T_a + n_R R T_a \tag{22.5}$$

where n_R is the number of moles of gas in the added volume at P_a. Substituting $P_a V_R$ for $n_R R T_a$ into equation (22.5) gives

$$P_1(V_c - V_p + V_R) = n_2 R T_a + P_a V_R \tag{22.6}$$

Combining equations (22.5) and (22.6) results in

$$P_1(V_c - V_p + V_R) = P_2(V_c - V_p) + P_a V_R \tag{22.7}$$

Solving equation (22.7) for the sample volume V_p and allowing P_a to be zero, since all measurements are made above ambient pressure, results in

$$V_p = V_c + \frac{V_R}{1 - P_2/P_1} \tag{cf.22.1}$$

The measurement of V_c and V_R is accomplished by two pressurizations, one with the sample cell empty ($V_p = 0$) and one with a calibrated blank or known volume in the sample cell. For an empty cell equation (22.1) becomes

$$V_p = 0 = V_c + \frac{V_R}{1 - P_2/P_1} \tag{22.8}$$

When a calibrated blank of known volume is used, its volume, V_{cal}, can be expressed as

$$V_p = V_{cal} = V_c + \frac{V_R}{1 - P_2'/P_1'} \tag{22.9}$$

Combining equations (22.8) and (22.9) and solving for V_R give

230 Density measurement

$$V_R = \frac{V_{cal}}{\left[\frac{1}{(P_2/P_1 - 1)} - \frac{1}{(P_2'/P' - 1)}\right]} \quad (22.10)$$

Substitution of this value for V_R into equation (22.8) gives the cell volume V_c for subsequent use in the working equation (22.1).

The above derivation assumes ideal gas behavior which is closely obeyed at pressures near ambient room temperature by both pure helium and nitrogen. However, helium is preferred because of its smaller size.

When measuring powder volume in the manner described above, it is necessary to avoid using any gas which can be even slightly adsorbed. If so much as a thousandth of a monolayer were adsorbed, the equivalent volume of gas would be in the order of 0.001 cm^3 for each 2.84 m^2 of area, if nitrogen were used. Since the sample cell used in the apparatus described in Fig. 22.1 can hold 130 cm^3, the total surface area of the sample can be hundreds or even thousands of square meters. Thus, errors of $0.1-1.0 \text{ cm}^3$ can be incurred due to very small amounts of adsorption. This is another reason helium is recommended in any gas pycnometer.

A second source of error encountered with high area powders, is the annulus volume which exists between the powder surface and the closest approach distance of a gas molecule. Assuming that the closest approach of the helium atom to the powder surface is 0.5Å or 5×10^{-11} m and that the specific area of the powder is $1000 \text{ m}^2/\text{g}$, there will exist an annulus volume of $5 \times 10^{-8} \text{ m}^3/\text{g}$ or $5 \times 10^{-2} \text{ cm}^3/\text{g}$. This represents a density error of 5% on materials with densities near unity. The use of gases with molecules larger than helium will exacerbate this error, again indicating the use of helium as the preferred gas.

22.2 APPARENT DENSITY

When the fluid displaced by powder does not penetrate all the pores, the measured density will be less than the true density. When densities are determined by liquid displacement an apparent density is obtained which can differ according to the liquid used because of their different capacities to penetrate small pores. Therefore, when reporting apparent density, the liquid used should also be reported.

22.3 BULK DENSITY

The volume occupied by the solid plus the volume of voids when divided into the powder mass yields the bulk density. Therefore, when powder is poured into a graduated container, the bulk density is the mass divided by the volume of the powder bed.

22.4 TAP DENSITY

The tap density is another form of bulk density obtained by tapping or vibrating the container in a specified manner to achieve more efficient particle packing. The tap density is therefore usually greater than the bulk density.

22.5 EFFECTIVE DENSITY

A particle may contain embedded foreign bodies which may increase or decrease its density. It may also contain blind pores which are totally encapsulated by the particle and will effectively reduce the particle's density. In this case, the measured quantity is the effective density.

Table 22.1 Porosimetry density measurement

Sample I.D. Silica Gel   Date _____

Operator _____   Outgassing Conditions _____

DATA
1. Weight of empty cell   35.9483 g
2. Weight of cell filled with mercury   87.1300 g
3. Weight of cell and sample   37.0265 g
4. Weight of cell and sample filled with mercury   73.1008 g
5. Intruded volume at 414 MPa   0.406 cm^3

CALCULATIONS (using mercury density table below)
6. Volume of mercury in cell without sample   3.778 cm^3
7. Volume of mercury in cell with sample   2.663 cm^3
8. Volume of sample and pores smaller than 7.26 µm (6 − 7)   1.115 cm^3
9. Volume of sample and pores smaller than 18 Å (8 − 5)   0.709 cm^3
10. Sample density $\dfrac{(3-1)}{9}$ 1.52 g/cm^3

Densities of Mercury

Temperature (°C)	Density (g/cm^3)	Temperature (°C)	Density (g/cm^3)
15	13.5585	23	13.5389
16	13.5561	24	13.5364
17	13.5536	25	13.5340
18	13.5512	26	13.5315
19	13.5487	27	13.5291
20	13.5462	28	13.5266
21	13.5438	29	13.5242
22	13.5413	30	13.5217

232 Density measurement

22.6 DENSITY BY MERCURY POROSIMETRY

Mercury porosimetry provides a convenient method for measuring the density of powders. This technique gives the true density of those powders which do not possess pores or voids smaller than those into which intrusion occurs at the highest pressure attainable in the porosimeter and provides apparent densities for those powders that have pores smaller than those corresponding to the highest pressure.

The worksheet shown in Table 22.1 illustrates the calculation of the density of a silica gel sample using the intruded volume at 414 MPa.

The volume of the sample, including pores smaller than 7.26 μm is first determined at ambient pressure (0.1 MPa).

This is accomplished by weighing the cell filled with mercury and then the cell containing sample filled with mercury. These weighings must be carried out with dilatometer stem filled to the same level. After converting the weights of mercury to the corresponding volumes, using the density table, the sample volume can be determined as the difference between the two mercury volumes.

The volume of the sample and thus, the density, including pores smaller

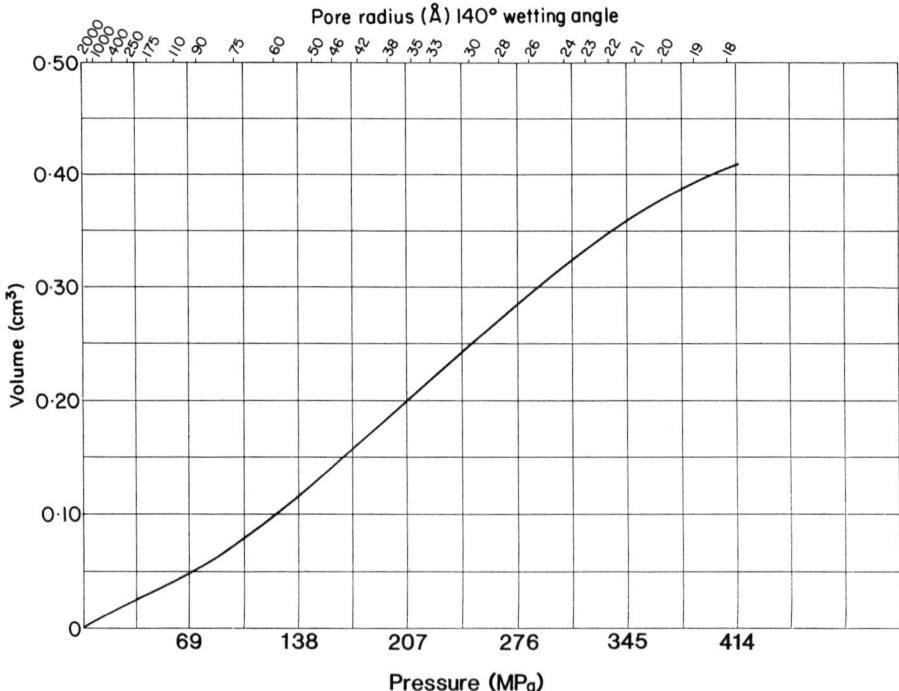

Figure 22.2 High-pressure intrusion curve of silica gel.

Density by mercury porosimetry 233

Table 22.2 Density of silica gel at various pressures

Pressure (MPa)	Radius* (Å)	Intruded volume† (cm³)	Sample volume (cm³)	Apparent density (g/cm³)
0.10	72,600	0	1.115‡	0.967
34	213	0.024	1.091	0.988
69	107	0.050	1.065	1.01
103	71.1	0.080	1.035	1.04
138	53.4	0.117	0.998	1.08
172	42.7	0.160	0.955	1.13
207	35.6	0.203	0.912	1.18
241	30.5	0.245	0.870	1.24
276	26.7	0.285	0.830	1.30
310	23.7	0.323	0.792	1.36
345	21.3	0.357	0.758	1.42
379	19.4	0.385	0.730	1.48
414	17.8	0.406	0.709	1.52

* Calculated from the intrusion pressure, assuming $\theta = 140°$; † for a sample weighing 1.078 g; ‡ taken from entry 8 of Table 22.1.

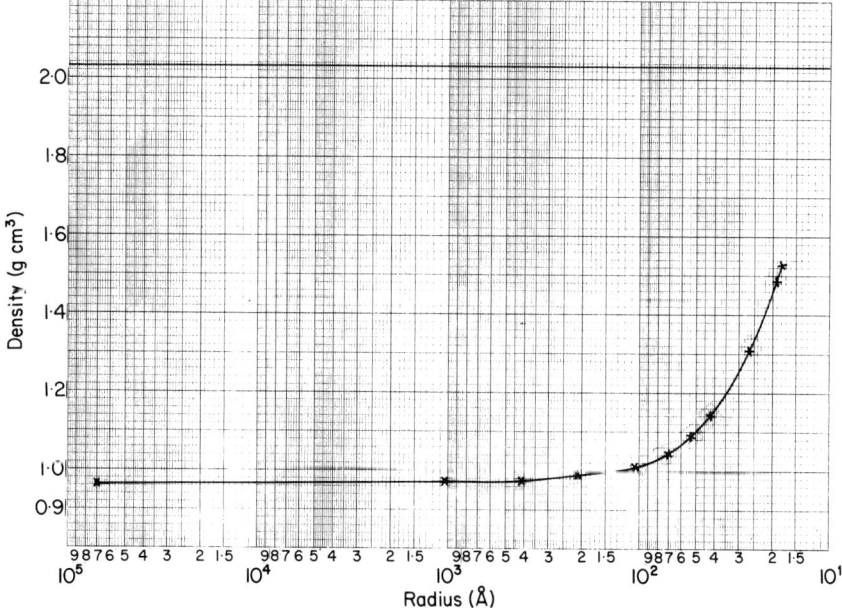

Figure 22.3 Density versus pore radius plot for silica gel.

than 18 Å, is calculated as the difference between the sample volume, shown as entry 8 in Table 22.1, and the volume of mercury intruded at 414 MPa.

The apparent density, that is, the volume of a given mass of sample plus voids divided into the sample mass can be calculated as a function of the void and pore volume from a mercury intrusion curve. The ambient to 414 MPa curve for the silica gel sample is illustrated in Fig. 22.2. Using the volume of mercury intruded at various pressures, the volume of the sample including voids and pores, and thus, the apparent density can be obtained, as shown in Table 22.2. The calculated apparent densities are obtained by subtracting the intruded volume from the initial sample volume and dividing the resulting value into the sample weight.

A plot of density versus pore radius, from the data in Table 22.2, is shown in Fig. 22.3. The horizontal line indicates the true density obtained by helium pycnometry. This higher density by gas displacement reflects the volume of pores smaller than about 18 Å.

23
Particle size analysis

23.1 INTRODUCTION

The particle size technique described in Chapter 13 combines gravitational sedimentation with the detection of X-ray absorption by a homogeneous suspension of particles. Fig. 23.1 is a schematic diagram of the X-ray sedimentometer shown in Fig. 23.2. The solid concentration is measured at the start of a sedimentation analysis and subsequently, depending upon the analysis parameters, the concentration is measured continuously with time and at various depths, h, below the surface. This is accomplished by scanning the sedimentation vessel with the X-ray beam from the bottom to the top of the sample cell. The hold time near the bottom of the cell and subsequent scan rate are determined by the analysis parameters, for example, maximum and minimum diameters chosen and minimum sedimentation distance.

Radiation from an X-ray tube (1) in Fig. 23.1 is collimated through slits (2) on both sides of the sample cell (3) containing the sedimenting suspension of particles. The radiation which is not absorbed by the particles is detected by a scintillation counter (4) and converted to a voltage which represents the intensity of the flux. The concentration of particles is subsequently calculated from the transmitted flux intensity as the percent mass finer than a specified equivalent spherical diameter.

Fig. 23.3 illustrates a typical particle size analysis, that is, a plot of the cumulative mass percentage finer than versus equivalent spherical diameter D.

23.2 SAMPLE PREPARATION

Accurate particle size determinations require that a powder of known density be well-dispersed in a liquid of known density and viscosity.

If a handbook value for the powder density is not available or is not sufficiently accurate, the true density, measured by helium pycnometry, can be used. If the material is thought to have closed pores, into which the sedimentation liquid cannot penetrate, the Stokes' density should be measured in the liquid to be used for the particle size analysis. This can be accomplished with a liquid pycnometer.

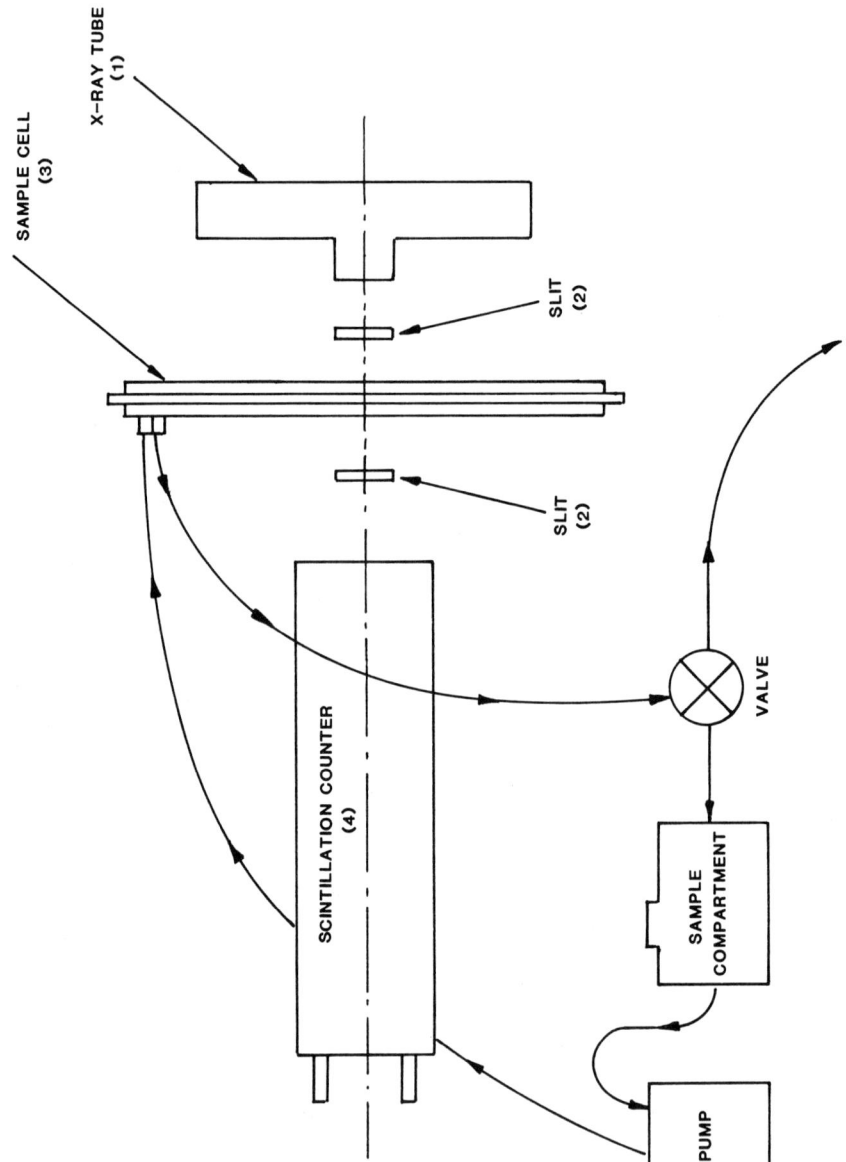

Figure 23.1 Schematic of X-ray sedimentometer.

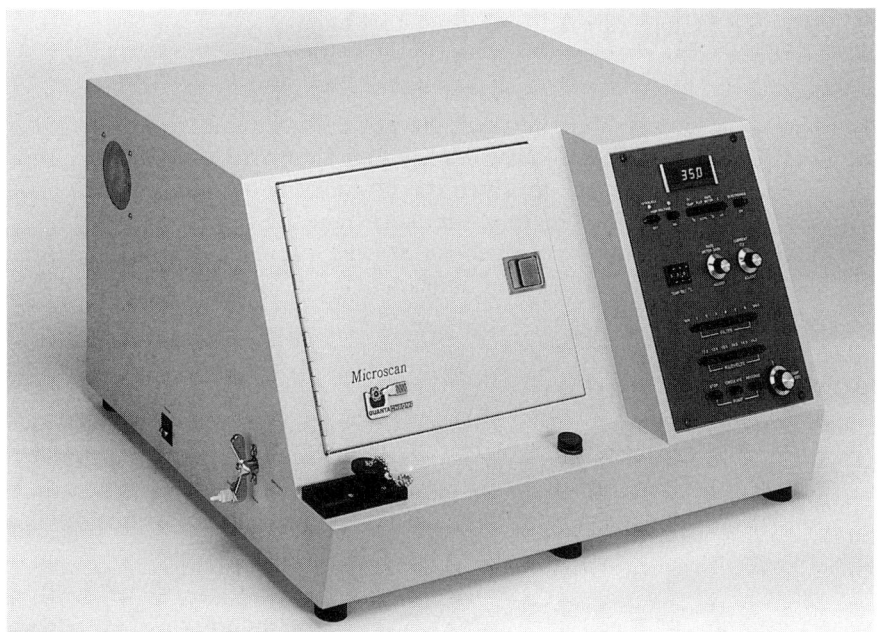

Figure 23.2 X-ray sedimentometer (courtesy of Quantachrome Corp.)

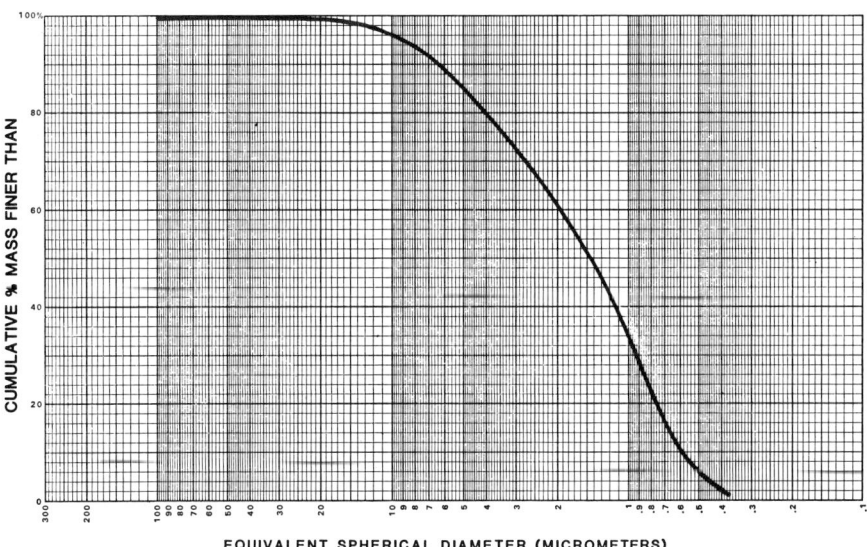

Figure 23.3 Particle size distribution.

23.3 SEDIMENTATION FLUID

The sedimentation liquid must fulfill the following requirements:

1. The liquid must be one in which the powder is insoluble.
2. The liquid must have a lower density than the particles to be analyzed.
3. The liquid must be one in which the powder can be well-dispersed.
4. The liquid should have a viscosity such that the largest particles to be analyzed will remain suspended for at least a few seconds and such that the liquid flow is laminar, as indicated by the Reynolds number (see Section 13.5).

23.4 DISPERSION

Prior to measuring a particle size distribution, it is essential that the particles be evenly dispersed in the sedimentation fluid. Dispersion of a powder consists of 3 steps:

Step 1: Wetting the powder

Wetting the powder, that is, establishing the solid–liquid interface, usually requires the addition of an appropriate dispersing agent (surfactant). For most powders, the typical concentration of wetting agent is 0.1 weight percentage by volume. A test for the wetting efficiency of dispersing agents is the observation of the settling of suspensions of approximately the same powder concentration but with different surfactants. A good dispersing agent will cause a slow settling of the particles with a clear interface between the liquid and powder and a small depth of sediment when settling is complete.

Ionic or polar nonionic dispersing agents can generally be used to disperse most powders. Metals and metal oxides readily disperse in a 0.1 weight % by volume of tetrasodium pyrophosphate (TSPP), Calgon (sodium hexametaphosphate) or Triton X-100 (octyl phenoxy polyethoxy ethanol). Sodium linoleate is recommended for the dispersion of nonpolar powders in aqueous media, while nonpolar particles in nonaqueous media can generally be dispersed with xylene. Table 23.1 lists a few of the most common dispersing agents along with some guidelines for their proper choice.

Step 2: Deagglomeration of particle clusters

For easily-wetted powders, the addition of a dispersing agent and the action of a circulation pump causes the particles to separate. However, in many cases, it is necessary to apply additional mechanical energy to the suspension. An ultrasonic probe inserted into the suspension during circulation is usually sufficient to separate any agglomerates.

To test for particle dispersion, a few drops of the suspension should be

Table 23.1

Dispersing agent	Type	Use
TSPP	ionic	for ionic particles in aqueous media
Calgon	ionic	for ionic particles in aqueous media
Sodium linoleate	ionic	for nonpolar particles in aqueous media
Triton X-100	nonionic (polar)	for ionic particles in aqueous media
Xylene	nonionic (nonpolar)	for nonpolar particles in nonaqueous media

viewed microscopically. A well-dispersed suspension will appear as discrete well-separated particles. In addition, if the particle size distribution shows smaller size particles when the analysis is rerun, the original suspension was not well-dispersed. Reanalysis of the suspension after dilution is another technique used to determine if particles were agglomerated.

Step 3: Stabilization of the dispersed suspension

Some dispersed suspensions, when allowed to stand for a short time, will flocculate, that is, the particles will agglomerate and settle en masse. Flocculation can easily be verified by the observation of a sudden sharp drop in mass, after the analysis has started.

23.5 DATA REDUCTION

The X-ray flux intensities measured during a sedimentation analysis can be expressed as cumulative percentage mass (M) finer than a specified diameter using equation (13.30)

$$\% M = \frac{\ln(I/I_f)}{\ln(I_{min}/I_f)} \times 100 \quad \text{(cf.13.30)}$$

where I is the flux at diameter D, I_{min} is the flux at the beginning of the analysis and I_f is the flux through pure fluid.

If M_N is assumed to be the mass of particles at diameter D_N and M_{N+1} the mass at diameter D_{N+1}, where D_N is greater than D_{N+1}, then S, the surface area per gram of total mass in the range D_N to D_{N+1} can be calculated

$$S = \frac{6(M_N - M_{N+1})}{\rho D_N} \quad (23.1)$$

assuming spherical particles of density, ρ.

240 Particle size analysis

Similarly, N, the number of particles per total mass in the size range D_N to D_{N+1} is given by

$$N = \frac{6(M_N - M_{N+1})}{\rho D^3} \quad (23.2)$$

The surface area and particle number values from equations (23.1) and (23.2) can be expressed as cumulative percentage functions using the following conversions.

$$\% \, S = \frac{\sum_{i=1}^{j} S_i - \sum_{i=1}^{N_i} S_i}{\sum_{i=1}^{j} S_i} \times 100 \quad (23.3)$$

and

$$\% \, N = \frac{\sum_{i=1}^{j} N_i - \sum_{i=1}^{N} N_i}{\sum_{i=1}^{j} N_i} \times 100 \quad (23.4)$$

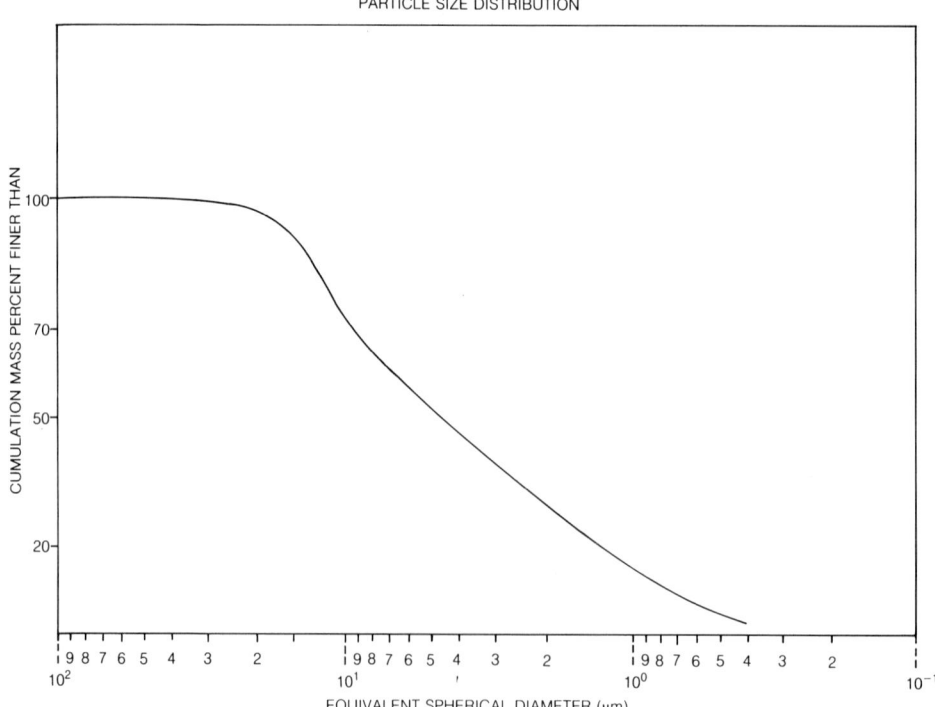

Figure 23.4 Cumulative mass distribution.

where S_i and N_i are the surface area/gram and number/gram in each interval and j is the total number of intervals.

The total surface area/gram (S_T) and total number of particles/gram (N_T) are given by

$$S_T = \sum_{i=1}^{j} S_i$$

$$N_T = \sum_{i=1}^{j} N_i$$

Typical cumulative mass, surface area and number distributions plotted versus equivalent spherical diameter are shown in Figs. 23.4–23.6.

A commonly used method of data presentation in particle sizing is statistical, that is, to report data as the median, mode and mean of a distribution plot. The median of a distribution is the 50% point; the mode is the maximum on the distribution derivative plot and the mean is defined as

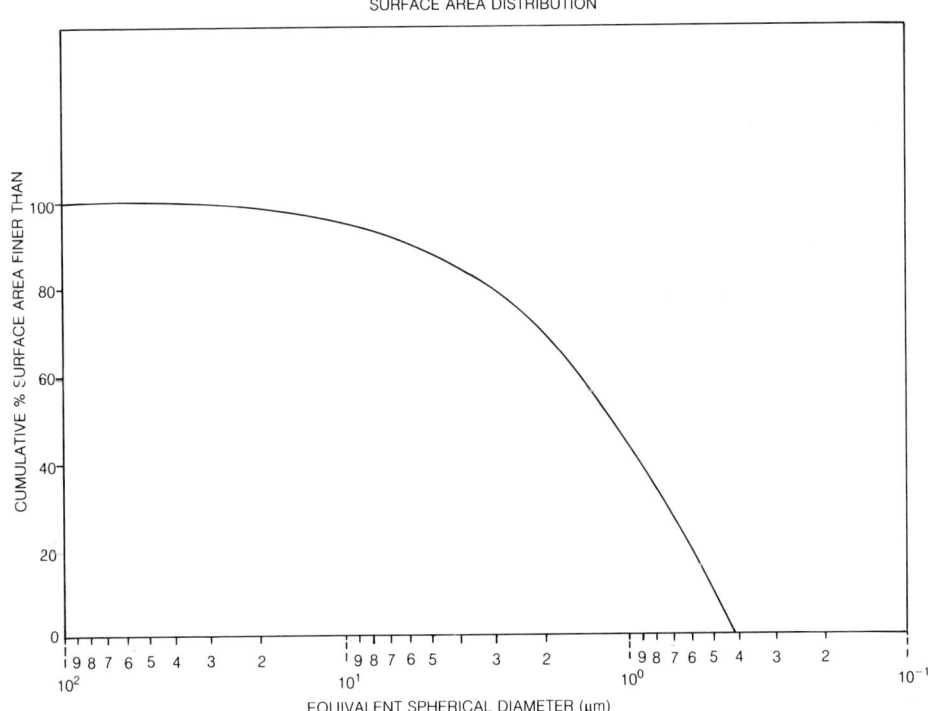

Figure 23.5 Cumulative surface area distribution.

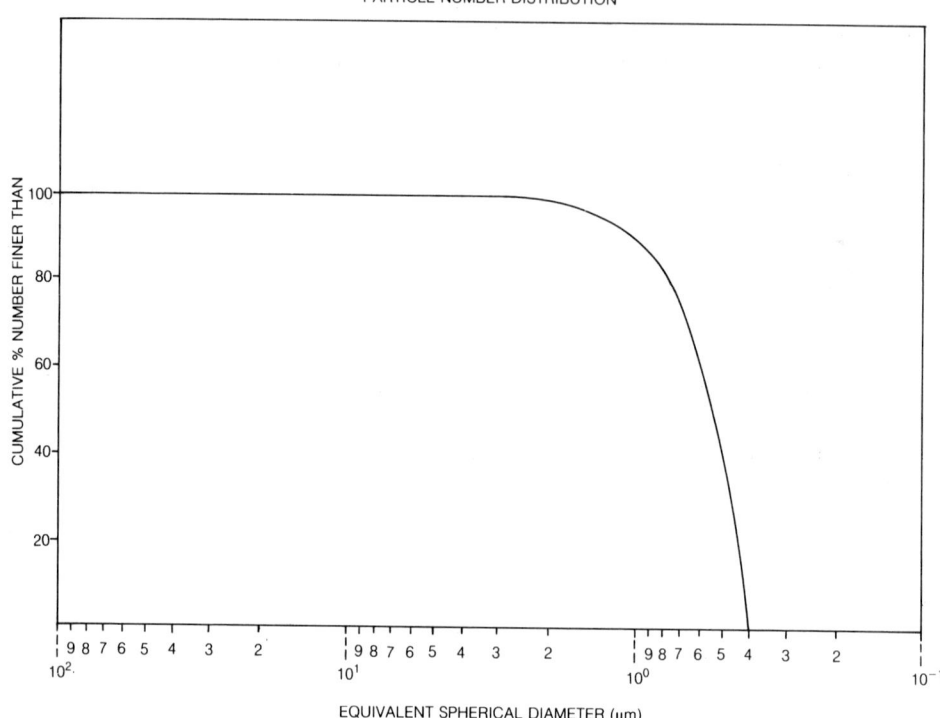

Figure 23.6 Cumulative number distribution.

$$\text{Mean} = \frac{\sum_{i=1}^{N}(\Delta M_i D_i)}{\sum_{i=1}^{N} \Delta M_i} \quad (23.5)$$

where ΔM is the change in mass in an interval and D_i is the average diameter in that interval.

References

1. Coulter Electronics, Hialeah, Florida.
2. Orr, Jr. C. and Dalla Valle, J. M. (1959) *Fine Particle Measurement*, Macmillan, New York.
3. Allen, T. (1981) *Particle Size Measurement*, Chapman and Hall, London.
4. London, F. (1930) *Z. Phys.*, **63**, 245.
5. Brunauer, S., Deming, L. S., Deming, W. S. and Teller, E. (1940) *J. Amer. Chem. Soc.*, **62**, 1723.
6. Langmuir, I. (1918) *J. Amer. Chem. Soc.*, **40**, 1368.
7. Brunauer, S., Emmett, P. H. and Teller, E. (1938) *J. Amer. Chem. Soc.*, **60**, 309.
8. Hill, T. L. (1946) *J. Chem. Phys.*, **14**, 268.
9. Lowell, S. (1975) *Powder Technol.*, **12**, 291.
10. Brunauer, S., Copeland, L. E. and Kantro, D. L. (1967) *The Gas Solid Interface*, Vol. 1, Dekker, New York, Ch. 3.
11. Brunauer, S. (1961) *Solid Surfaces and the Gas Solid Interface*, Advances in Chemistry Series, No. 33, American Chemical Society, Washington, D.C.
12. MacIver, D. S. and Emmett, P. H. (1956) *J. Amer. Chem. Soc.*, **60**, 824.
13. Sandstede, G. and Robins, E. (1960) *Chem. Ing. Tech.*, **32**, 413.
14. Gregg, S. J. and Sing, K. S. W. (1982) *Adsorption, Surface Area and Porosity*, 2nd edn, Academic Press, New York.
15. Emmett, P. H. and Brunauer, S. (1937) *J. Amer. Chem. Soc.*, **59**, 1553.
16. Moelwyn-Hughes, E. A. (1961) *Physical Chemistry*, 2nd edn, p. 545, Pergamon Press, New York.
17. Kiselev, A. V. and Eltekov, Y. A. (1957) *International Congress on Surface Activity* (1957) II, Butterworths, London, p. 228.
18. Lowell, S., Shields, J. E., Charalambous, G. and Manzione, J. (1982) *J. Colloid Interf. Sci.*, **86**, 191.
19. Harris, B. L. and Emmett, P. H. (1949) *J. Phys. Chem.*, **53**, 811.
20. Livingston, H. K. (1949) *J. Colloid Sci.*, **4**, 1447.
21. Rhodin, T. N. (1953) *J. Phys. Chem.*, **57**, 1437.
22. Isirikyan, A. A. and Kiselev, A. V. (1961) *J. Phys. Chem.*, **65**, 601; *ibid.*, **66**, 210.
23. O'Conner, T. L. and Uhlig, H. H. (1957) *J. Phys. Chem.*, **61**, 402.
24. Zettlemoyer, A. C., Chand, A. and Gamble, E. (1950) *J. Amer. Chem. Soc.*, **72**, 2752.
25. Davis, R. T., DeWitt, T. W. and Emmett, P. H. (1947) *J. Phys. Chem.*, **51**, 1232.
26. Emmett, P. H. and Brunauer, S. (1937) *J. Amer. Chem. Soc.*, **59**, 1553.
27. Brunauer, S. and Emmett, P. H. (1937) *J. Amer. Chem. Soc.*, **59**, 2682.
28. Chenebault, P. and Schurenkampfer, A. (1965) *J. Phys. Chem.*, **69**, 2300.
29. Harkins, W. D. and Jura, G. (1944) *J. Amer. Chem. Soc.*, **66**, 919.

30. Langmuir, I. (1917) *J. Amer. Chem. Soc.*, **39**, 1848.
31. Adamson, A. W. (1982) *Physical Chemistry of Surfaces*, 4th edn, Wiley Interscience, New York, Ch. 3.
32. Gibbs, J. W. (1931) *The Collected Works of J. W. Gibbs*, Vol. 1, Longmans Green, New York.
33. Guggenheim, E. A. and Adam, N. K. (1936) *Proc. Roy. Soc. London*, **154A**, 608.
34. Emmett, P. H. (1946) *J. Amer. Chem. Soc.*, **68**, 1784.
35. Harkins, W. D. and Jura, G. (1944) *J. Amer. Chem. Soc.*, **66**, 1362.
36. Gregg, S. J. and Sing, K. S. W. (1967) *Adsorption, Surface Area and Porosity*, Academic Press, New York, p. 301.
37. Poiseuille, J. L. M. (1846) *Inst. France Acad. Sci.*, **9**, 433.
38. Darcy, H. P. G. (1856) *Les Fontaines Publiques de la Ville de Dijon*, Victor De Lamont.
39. Kozeny, J. (1927) *Ber. Wien Akad.*, **136A**, 271.
40. Carmen, P. C. (1938) *J. Soc. Chem. Ind., London, Trans. Commun.*, **57**, 225.
41. Ridgen, P. J. (1954) *Road Res. Pap.*, No. 28 (NMSO).
42. Lea, F. M. and Nurse, R. W. (1947) *Trans. Inst. Chem. Eng.*, **25**, 47.
43. Carman, P. C. (1956) *Flow of Gases Through Porous Media*, Butterworths, London.
44. Allen, T. (1990) *Particle Size Measurement*, 4th edn, Chapman and Hall, London, p. 615.
45. Thompson, W. T. (1871) *Philos. Mag.*, **42**, 448.
46. Zsigmondy, A. (1911) *Z. Anorg. Chem.*, **71**, 356.
47. McBain, V. W. (1935) *J. Amer. Chem. Soc.*, **57**, 699.
48. Cohan, L. H. (1938) *J. Amer. Chem. Soc.*, **60**, 433.
49. Foster, A. G. (1952) *J. Chem. Soc.*, 1806.
50. Broekhoff, J. C. P. and Linsen, B. G. (1970) *Physical and Chemical Aspects of Adsorbents and Catalysts*, (ed. B. G. Linsen), Academic Press, New York, Ch. 1.
51. de Boer, J. H. (1958) *The Structure and Properties of Porous Materials*, Butterworths, London, p. 68.
52. Broad, D. W. and Foster, A. G. (1946) *J. Chem. Soc.*, 447.
53. Foster, A. G. (1934) *Proc. Roy. Soc. London*, **147A**, 128.
54. McKee, D. W. (1959) *J. Phys. Chem.*, **63**, 1256.
55. Brown, M. J. and Foster, A. G. (1951) *J. Chem. Soc.*, 1139.
56. Gurvitsh, L. (1915) *J. Phys. Chem. Soc. Russ.*, **47**, 805.
57. Lippens, B. C., Linsen, B. G. and de Boer, J. H. (1964) *J. Catal.*, **3**, 32.
58. Shull, C. G. (1958) *J. Amer. Chem. Soc.*, **70**, 1405.
59. Lippens, B. C. and de Boer, J. H. (1965) *J. Catal.*, **4**, 319.
60. Pierce, C. (1953) *J. Phys. Chem.*, **57**, 149.
61. Harris, M. R. and Sing, K. S. W. (1959) *Chem. Ind. (London)*, 11.
62. Cranston, R. W. and Inkley, F. A. (1957) *Adv. Catal.*, **9**, 143.
63. Halsey, G. D. (1948) *J. Chem. Phys.*, **16**, 931.
64. Orr, Jr., C. and Dalla Valle, J. M. (1959) *Fine Particle Measurement*, Macmillan, New York, Ch. 10.
65. Barrett, E. P., Joyner, L. G. and Halenda, P. P. (1951) *J. Amer. Chem. Soc.*, **73**, 373.
66. Wheeler, A. (1945, 1946) Catalyst Symposia, Gibson Island A.A.A.S. Conference.
67. Pierce, C. and Smith, R. N. (1953) *J. Phys. Chem.*, **57**, 64.
68. Carman, P. C. (1953) *J. Phys. Chem.*, **57**, 56.

69. Oulton, T. D. (1948) *J. Phys. Colloid Chem.*, **52**, 1296.
70. Roberts, B. F. (1963) *A Procedure For Estimating Pore Volume and Area Distributions from Sorption Isotherms*, National Meeting of the American Chemical Society.
71. Innes, W. B. (1957) *Anal. Chem.*, **29**, 1069.
72. Brunauer, S., Mikhail, R. S. and Boder, E. E. (1967) *J. Colloid Interface Sci.*, **24**, 451.
73. Kiselev, A. V. (1945) *Usp. Khim.*, **14**, 367.
74. Dubinin, M. M. (1960) *Zh. Fiz. Khim.*, **34**, 959.
75. van der Plas, Th. (1970) Physical and Chemical Aspects of Adsorbents and Catalysts (ed. B. G. Linsen), Academic Press, New York, Ch. 9.
76. Polanyi, M. (1914) *Verh. Dtsch. Phys. Ges.*, **16**, 1012.
77. Dubinen, M. M. (1960) *Chem. Rev.*, **60**, 235.
78. Dubinen, M. M. and Timofeev, D. P. (1948) *Zh. Fiz. Khim.*, **22**, 133.
79. Dubinen, M. M. and Radushkevich, L. V. (1947) *Dokl. Akad. Nauk. SSSR*, **55**, 331.
80. Nikolayev, K. M. and Dubinen, M. M. (1958) *Izv. Akad. Nauk. SSSR, Otd. Tekn. Nauk.*, 1165.
81. Kaganer, M. G. (1959) *Zh. Fiz. Khim.*, **32**, 2209.
82. de Boer, J. H., Lippens, B. C., Linsen, B. G., Broekhoff, J. C. P., van den Heuvel, A., and Osinga, T. V. (1966) *J. Colloid Interf. Sci.*, **21**, 405.
83. Johnson, M. F. L. (1978) *J. Catalysis*, **52**, 425.
84. Gregg, S. J. and Sing, K. S. W. (1982) *Adsorption, Surface Area and Porosity*, 2nd edn, Academic Press, New York.
85. Mikhail, R. H., Brunauer, S. and Bodor, E. E. (1968) *J. Colloid Interface Sci.*, **26**, 45.
86. Brunauer, S. (1972) *J. Colloid Interface Sci.*, **39**, 435; **41**, 612.
87. de Boer, J. H., Linsen, B. G., van der Plas, T. and Zondervan, G. J. (1965) *J. Catal.*, **4**, 649.
88. Shields, J. E. and Lowell, S. (1983) *Powder Technol.*, **36**, 1.
89. Young, T. (1855) *Miscellaneous Works*, (ed. G. Peacock), Vol. 1, J. Murray, London, p. 418.
90. Laplace, P. S. (1806) *Mechanique Celeste*, Supplement to Book 10.
91. Dupre, A. (1869) *Theorie Mecanique de la Chaleur*, p. 368.
92. Washburn, E. W. (1921) *Phys. Rev.*, **17**, 273.
93. Ritter, L. C. and Drake, R. L. (1945) *Ind. Eng. Chem. Anal. Ed.*, **782**, 17.
94. Frevel, L. K. and Kressley, L. J. (1963) *Anal. Chem.*, **35**, 1492.
95. Stanley-Wood, N. G. (1979) *Analyst*, **104**, 79.
96. Lee, J. A. and Maskell, W. C. (1974) *Powder Technol.*, **9**, 165.
97. Mayer, R. P. and Stowe, R. A. (1966) *J. Phys. Chem.*, **70**, 3867.
98. Mayer, R. P. and Stowe, R. A. (1965) *J. Colloid Interface Sci.*, **20**, 893.
99. Lowell, S. and Shields, J. E. (1981) *Powder Technol.*, **28**, 201.
100. Quantachrome Corporation, Syosset, New York.
101. Zwietering, P. (1958) *The Structure and Properties of Porous Solids*, Butterworths, London, p. 287.
102. Joyner, L. G., Barrett, E. P. and Skold, R. (1951) *J. Amer. Chem. Soc.*, **73**, 3158.
103. Cochran, C. N. and Cosgrove, L. A. (1957) *J. Phys. Chem.*, **61**, 1417.
104. Dubinen, M. M., Vishnyakova, E. G., Zhukovskaya, E. A., Leontev, E. A., Lukyanovich, V. M. and Saraknov, A. I. (1960) *Russ. J. Phys. Chem.*, **34**, 959.
105. Rootare, H. M. and Prenzlow, C. F. (1967) *J. Phys. Chem.*, **71**, 2733.
106. Henrion, P. N., Gienen, F. and Leurs, A. (1977) *Powder Technol.*, **16**, 167.

References

107. Lowell, S. and Shields, J. E. (1981) *J. Colloid Interface Sci.*, **80**, 192.
108. Lowell, S. and Shields, J. E. (1981) *J. Colloid Interface Sci.*, **83**, 273.
109. Lowell, S. and Shields, J. E. (1982) *J. Colloid Interface Sci.*, **90**, 203.
110. Gregg, S. J. and Sing, K. S. W. (1982) *Adsorption, Surface Area and Porosity*, 2nd edn, Academic Press, New York, p. 195.
111. Adamson, A. W. (1982) *Physical Chemistry of Surfaces*, 4th edn, Wiley Interscience, New York, p. 242.
112. Polanyi, M. (1914) *Verh. Dtsch. Phys. Ges.*, **16**, 1012.
113. Orr, Jr. C. (1970) *Powder Technol.*, **3**, 117.
114. Lowell, S. (1980) *Powder Technol.*, **25**, 37.
115. Reverberi, G., Feraiolo, G. and Peloso, A. (1966) *Ann. Chim. (Italy)*, **56**, 1552.
116. Androutsopoulos, G. P. and Mann, R. (1979) *Chem. Eng. Sci.*, **34**, 1203.
117. Lowell, S. and Shields, J. E. (1981) *Powder Technol.*, **29**, 225.
118. Glasstone, S. (1946) *Textbook of Physical Chemistry*, 2nd edn, Van Nostrand, New York, p. 450.
119. Hatton, T. A. (1978) *Powder Technol.*, **19**, 227.
120. Lopez-Gonzales, J. de D., Carpenter, F. G. and Deitz, V. R. (1955) *J. Res. Nat. Bur. Stand.*, **55**, 11.
121. Orr, Jr., C. and Dalla Valle, J. M. (1959) *Fine Particle Measurement*, Macmillan, New York, p. 176.
122. Faeth, P. A. (1962) *Adsorption and Vacuum Technique*, Institute of Science and Technology, University of Michigan, Ann Arbor.
123. Emmett, P. H. (1941) A.S.T.M. *Symposium on New Methods for Particle Size Determination in the Sub-Sieve Range*, p. 95.
124. Harris, M. R. and Sing, K. S. W. (1955) *J. Appl. Chem.*, **5**, 223.
125. Emmett, P. H. and Brunauer, S. (1937) *J. Amer. Chem. Soc.*, **59**, 1553.
126. Benson, S. W. (1960) *The Foundation of Chemical Kinetics*, McGraw-Hill, New York, p. 66.
127. Maggs, F. A. P. (1953) *Research (London)*, **6**, 513.
128. Wynne-Jones, W. F. K. and Marsh, H. (1964), *Carbon*, **1**, 281.
129. Maggs, F. A. P. (1952) *Nature (London)*, **169**, 259; **169**, 793.
130. Zweitering, P. and van Krevelin, D. W. (1954) *Fuel*, **33**, 331.
131. Beebe, R. A., Beckwith, J. B. and Honig, J. M. (1945) *J. Amer. Chem. Soc.*, **67**, 1554.
132. Litvan, G. G. (1972) *J. Phys. Chem.*, **76**, 2584.
133. Edmonds, T. and Hobson, J. P. (1965) *J. Vac. Sci. Technol.*, **2**, 182.
134. Rosenberg, A. J. (1956) *J. Amer. Chem. Soc.*, **78**, 2929.
135. Shields, J. E. and Lowell, S. (1984) *Amer. Lab.*, Nov.
136. Northrop, P. S., Flagan, R. C., and Gavalas, G. R. (1987) *Langmuir*, **3**, 300.
137. Loebenstein, W. V. and Deitz, V. R. (1951) *J. Res. Nat. Bur. Stand.*, **46**, 51.
138. de Boer, J. H. (1953) *The Dynamical Character of Adsorption*, Oxford University Press, London, p. 33.
139. Nelson, F. M. and Eggertsen, F. T. (1958) *Anal. Chem.*, **30**, 1387.
140. Karp, S. and Lowell, S. (1971) *Anal. Chem.*, **43**, 1910.
141. Karp, S., Lowell, S. and Mustaccuiolo, A. (1972) *Anal. Chem.*, **44**, 2395.
142. Kourilova, D. and Krevel, M. (1972) *J. Chromat.*, **65**, 71.
143. Lowell, S. and Karp, S. (1972) *Anal. Chem.*, **44**, 1706.
144. Lowell, S. (1973) *Anal. Chem.*, **45**, 8.
145. Benson, S. W. (1960) *The Foundation of Chemical Kinetics*, McGraw-Hill, New York, p. 188.
146. Quantachrome Corp., Syosset, New York.
147. Haley, A. J. (1963) *J. Appl. Chem.*, **13**, 392.

148. Semonian, B. D. and Manes, M. (1977) *Anal. Chem.*, **49**, 991.
149. Wilson, J. N. (1940) *J. Amer. Chem. Soc.*, **62**, 1583.
150. Glueckauf, E. (1946) *Proc. Roy. Soc. London*, **186A**, 35.
151. Glueckauf, E. (1947) *J. Chem. Soc.*, 1302, 1308, 1315, 1327.
152. Glueckauf, E. (1949) *Discuss. Faraday Soc.*, **7**, 12.
153. De Vault, D. (1943) *J. Amer. Chem. Soc.*, **65**, 532.
154. Weiss, J. (1943) *J. Chem. Soc.*, 297.
155. Offord, A. C. and Weiss, J. (1945) *Nature (London)*, **155**, 725.
156. Stock, R. (1955) *Ph.D. Thesis*, London University, London.
157. Gregg, S. J. (1961) *The Surface Chemistry of Solids*, Reinhold, New York.
158. Cremer, E. E. and Huber, H. F. (1962) (eds N. Breuner, J. E. Callen and M. D. Weiss), Academic Press, New York, p. 169.
159. Malamud, H., Geisman, H. and Lowell, S. (1967) *Anal. Chem.*, **39**, 1468.
160. Cahn, L. and Schutz, H. R. (1962) *Vac. Microbalance Tech.*, **3**, 29.
161. McBain, J. and Bakr, A. M. (1926) *J. Amer. Chem. Soc.*, **48**, 690.
162. Young, D. M. and Crowel, A. D. (1962) *Physical Adsorption of Gases*, Butterworths, London.
163. Glasstone, S., Laidler, K. J. and Eyring, H. (1941) *The Theory of Rate Processes*, McGraw-Hill, New York.
164. Rhodes, J. F. and Katz, S. (1974) Meeting of the American Ceramic Society.
165. Gruber, H. L. (1962) *Anal. Chem.*, **34**, 1828.
166. Penn, L. S. and Miller, B. (1980) *J. Colloid Interface Sci.*, **77**, 574.
167. Shields, J. E. and Lowell, S. (1982) *Powder Technol.*, **31**, 227.
168. Winslow, D. N. (1978) *J. Colloid Interface Sci.*, **67**, 42.
169. Osipow, L. I. (1964) *Surface Chemistry*, Reinhold, New York, p. 233.

Index

α_s-method, 84
Activated physical adsorption, 168, 212
Activation energy, 212
Adsorbate interactions, 9, 28, 38
Adsorbed film depth, 59–64, 69
Adsorption
  chemical, 8, 210–16
  energy of, 15, 16
  epitaxial, 38
  heat of, 11, 28, 45
  hysteresis, 55
  hysteresis loop scan, 188
  hysteresis types, 56–8
  localized, 38
  multilayer, 18
  physical, 8, 17
  potential, 17, 28, 168, 174
Adsorption isotherms, 8, 11–13, 25
  Type I, 11, 14–17, 25, 54, 72, 201
  Type II, 11, 25, 54, 201
  Type III, 11, 25, 27, 54
  Type IV, 11, 25, 54, 201
  Type V, 12, 25, 54, 201
  Type VI, 13
Affinity coefficient, 74
Agglomeration, 238
Amalgamation, 129
Approach velocity, 48
Argon adsorption, 169
Arrhenius equation, 168
Aspect factor, 50

BET C constant, 23, 86, 170
  single point, 30–4
BET equation, 22
BET theory, 17–22
Bottle-neck pores, 56, 58, 59, 61, 130

Brownian Motion, 144–50
Bouyancy, 202–4, 206

Capillarity, 90, 94–6
Channel diameter, 47
Channel length, 48
Chemisorption, 8, 14, 210–16
  heat of, 211
  isotherms, 212
Clapeyron–Clausius equation, 156
Coefficient of thermal diffusion, 192
Compressibility, 104
Condensation Coefficient, 15
Contact Angle, 52–4, 90, 93–4, 99
Continuous flow gas adsorption
  apparatus, 174, 175
  data reduction, 193
  sample cells, 184
Critical temperature, 73, 174
Cross-sectional area, 14, 17, 35–41, 60
  and C constant, 38–40
  table of, 41

Darcy's law, 47
Degassing, 160, 165, 208
Density, powder, 50
  apparent, 230
  bulk, 157, 230
  by helium pycnometry, 227–30
  by mercury porosimetry, 232
  effective, 231
  tap, 231
  true, 227
Desorption isotherm, 55, 59
Diffusion, 47, 144–50
Diffusional flow, 47, 50
Dilatometer, 217, 218

Dispersion, 238
Dynamic methods, 174–96

Effective cross-sectional area, 14
Electron microscopy, 6
Envelope surface area, 47, 51
Equilibrium vapor pressure, 52, 170
Equipotential plane, 73
Equivalent spherical diameter, 136, 239

Fick's Law, 146
Film balance, 42
Fractional Coverage, 24
Free energy, 52, 55
Free surface energy, 90
Frontal analysis, 198–201

Gas adsorption, 7–10
Gibbs' adsorption equation, 42
Gibbs' free energy, 9
Gravimetric methods, 202–4
Gurvitsh rule, 58

Halsey equation, 61, 68
Harkins–Jura method, 44–6
Harkins–Jura constant, 45
Heat of adsorption, 11, 28, 45
Heat of chemisorption, 210, 212
Heat of condensation, 45
Heat of immersion, 45
Heat of liquifaction, 11, 18, 28
Helical spring balance, 204
Helium, 174
Helium pycnometry, 228–30
Henry's law, 211
Hess's law, 45
Hydraulic radius, 67, 88
Hysteresis, adsorption, 55–6, 188
    energy, 127
    mercury intrusion–extrusion, 104, 121, 126
    types, 56–8

Ideal gas corrections, 165, 206
Interaction potential, 7
Interparticle voids, 100
Intrusion–extrusion curves, 100–3
    common features, 103

Isotherm, adsorption, 8, 11–13, 54, 188
    continuous, 198
    desorption, 55–6, 59, 164, 188

Kelvin equation, 52–4, 57, 65, 119, 131
Kelvin radius, 59, 61
Kozeny equation, 47
Krypton adsorption, 169

Langmuir equation, 16, 72, 211
Langmuir isotherm, 15, 72
Latent heat of condensation, 45
Lateral interactions, 27
Linear flow velocity, 49
Liquid molar volume, 35, 53, 61
Localized adsorption, 38
Low surface area, 168–70, 189–93

Macropores, 72
Mean free path, 46, 170
Mercury contact angle, 223–6
    table of, 226
Mercury Porosimetry, 121–3, 223
    contact angle in, 121, 223
    data reduction, 99–120, 223
    hysteresis in, 104, 121–34
    isotherms, 131
    low pressure, 222
    pore length distribution from, 111
    pore population from, 111
    pore size distribution, 107–9
    pore surface area distribution from, 110
    scanning, 103
    surface area from, 218–23
    theory of, 90–8, 126
    volume distribution from, 107–9
    volume ln radius distribution from, 110
Mesopores, 72, 85
Microbalances, 202–5
Micropores, 11, 72–89
Micropore surface area
    from gas adsorption, 72–89
    from mercury porosimetry, 88
Micropore volume, 88
Modelless pore size analysis, 65
Molecular packing, 35–7

Monolayer, 16, 23, 27, 60
Micropore analysis method, 85–8
Multipoint method, 22, 175–7, 195

Nitrogen adsorption, 38, 59, 162, 187
   cross-sectional area, 35–41
   saturated vapor pressure, 170, 184
Non-wetting liquid, 93–5

Overlapping potential, 17, 52, 86

Particle porosity, 236
Particle size, 3, 5, 135–52, 238
Penetrometer, 99, 217
Permeametry, 5, 46–51
Poiseiulles' law, 47, 51
Polanyi potential theory, 73, 127, 198
Polarization forces, 29
Pore area, 105
Pore length, 111
Pore potential, 127
Pore radius, 59, 62, 64, 97, 107–9
Pore shape, 56, 59
Pore size, 52, 57, 64, 97, 107–9
Pore size distribution, 52, 57, 59–65, 107–9
Pore surface distribution, 67, 110
Pore volume, 58, 99
Porogram, 100
Porosimetric work, 124–5
Porosity, 4, 48
Powder volume, 227–30
Pressure jump method, 197
Pycnometer, 227–30

Reference standards, 155
Relative pressure, 25
Repetitive cycling, 160
Representative sampling, 156
Reynolds Number, 141
Rifflers, 156–60

Sample cell, 165, 184
Sample conditioning, 160, 165

Saturated vapor pressure, 170
Sedimentation, 135, 236
Sieving, 159
Single point method, 30–4, 194, 196
Site occupancy, 24
Slip flow, 47
Spreading pressure, 42
Standard
   reference, 155
Statistical thickness, 59, 69, 77
Stokes' law, 5, 135
Surface area
   envelope, 47, 50
   from BET equation, 22
   from mercury porosimetry, 105
   specific, 5, 23, 35
   total, 17, 23, 31
Surface interaction, 27
Surface pressure, 42
Surface tension, 42, 52, 65, 90, 94
Surface titration, 215

t-curve, 68, 70, 77
t-method, 77
Terminal velocity, 150–2
Thermal conductivity, 160, 175–83, 189
Thermal diffusion, 189–93
   coefficient of, 192
Thermal transpiration, 169, 192, 204
Thermomolecular flow, 204

Vacuum volumetric method, 162–73
Vapor, definition, 7
Viscous drag, 141
Void volume, 48, 157, 206
V–t curves, 54

Wall effects, 144
Washburn equation, 96–8, 108, 132
Wetting liquid, 93

X-ray absorption, 137–8, 239

Young and Laplace equation, 91